Code of

Estimating Practice

Code of
Estimating Practice

Seventh Edition

The Chartered Institute of Building

CIOB

WILEY-BLACKWELL

A John Wiley & Sons, Ltd., Publication

This edition first published 2009
© 2009 The Chartered Institute of Building

Blackwell Publishing was acquired by John Wiley & Sons in February 2007. Blackwell's publishing
programme has been merged with Wiley's global Scientific, Technical, and Medical business to
form Wiley-Blackwell.

Registered office
John Wiley & Sons Ltd, The Atrium, Southern Gate, Chichester, West Sussex, PO19 8SQ,
United Kingdom

Editorial offices
9600 Garsington Road, Oxford, OX4 2DQ, United Kingdom
2121 State Avenue, Ames, Iowa 50014-8300, USA

Fifth edition 1983
Sixth edition 1997
Seventh edition 2009

For details of our global editorial offices, for customer services and for information about how
to apply for permission to reuse the copyright material in this book, please see our website at
www.wiley.com/wiley-blackwell.

Library of Congress Cataloging-in-Publication Data
Code of estimating practice. – 7th ed.
  p. cm.
  Includes bibliographical references and index.
  ISBN 978-1-4051-2971-8 (pbk. : alk. paper)   1. Building—Estimates—Standards—
Great Britain.   2. Construction industry—Great Britain. I. Chartered Institute of
Building (Great Britain)

  TH435.C654 2009
  692′.5021841—dc22

                                                             2008053125

A catalogue record for this book is available from the British Library.

Set in Gothic 720 BT by Gray Publishing, Tunbridge Wells, Kent
Printed and bound in Malaysia by KHL Printing Co Sdn Bhd

2 2009

Contents

Contents

Preface to the Seventh Edition

The first edition of the *Code of Estimating Practice* was produced by the Institute's Estimating Practice Committee in April 1966. It has been revised on five occasions since that time, the last edition being published in 1997.

This edition of the Code has been completely re-structured and rewritten where necessary to reflect the changing pressures placed on the bid management and estimating processes. Rapidly changing procurement approaches have led to an ever-increasing range of tender documents needing to be understood and priced. These all carry varying risks for the contractor which need to be managed. Regulatory changes have taken place in recent years that have had a significant impact on the bid management and tendering process and include:

- the requirements of health, safety, welfare and environmental legislation

- company governance requirements for effective risk management.

Estimators have previously addressed these issues directly through the pricing of resources and costs of health, safety and welfare in preliminaries and risk-management measures built into estimating procedures. The requirements are now more demanding, with a specific need to demonstrate compliance and, therefore, these issues are addressed in the revised procedures.

In this update of the Code of Practice, sections on e-commerce and management of electronic information have been added.

The last edition was accompanied by four supplements. These have been incorporated where possible by editing sections, omitting outdated elements and simplifying procedures to reflect current best practice. Some of the checklists have been omitted as they were considered to be unnecessary and standard forms have been included in appendices within the relevant sections.

As with all best practice, management systems develop organically in response to changing needs. This Code endeavours to continue the excellent work of previous editions and to respond to current trends as reflected in best practice as observed by the team of contributors.

Good practice builds successful companies and enables successful tendering.

Acknowledgements

This new edition was compiled by a working group of representatives across the industry and edited by Sandra Lee, and Peter Darling BEng (Hons), CEng with Arnab Mukherjee BEng (Hons), MSc, MBA, MCIOB, MAPM providing additional support with technical editing. The composition of the working group was:

Saleem Akram FCIOB
Ian Bain
Martin Brook FCIOB
Michael Byng
Colin Duke
Rosemary Elder
Sandra Lee MCIOB
Nick Marsh MCIOB
Bill Massey FCIOB
Rob Mellor ICIOB
Sarah Peace
Terry Reiner MCIOB
Simon Rhoden
Harry Sharp FCIOB
Mike Shotter MCIOB
Martyn Stubbs MCIOB

The working group would also like to thank the following for commenting on the final draft:

Huntley Cartwright
Michael Dallas
Keith Pickavance FCIOB

Terminology

The following are the meanings of the principal specialist terms used in this Code.

All-in labour rate   A rate which includes payments to operatives and associated costs which arise directly from the employment of labour.

All-in material rate   A rate which includes the cost of material delivered to site, conversion, waste, unloading, handling, storage and preparing for use.

All-in mechanical plant rate   A rate which includes the costs originating from the ownership or hire of plant together with operating costs.

Approved contractors   Those who have demonstrated that they have the expertise, resources, ability and desire to tender for a proposed project. Selection of such contractors is normally by pre-selection procedures.

Attendance   The labour, plant, materials and/or other facilities provided by main contractors for the benefit of sub-contractors, for which sub-contractors normally bear no cost.

Awarding authority   The public sector body (department agency, NHS trust, local authority, etc.) which is procuring a service through the private finance initiative (PFI).

Benchmarking   A procedure for testing whether the standard and price of services is consistent with the market standard, without any formal competitive tendering. This is usually adopted during the project concession period to ensure faculty management services continue to represent value for money.

Buildability   The extent to which the design of a building facilitates ease of construction subject to the overall requirements of a completed building.

CDM Regulations   Construction (Design and Management) Regulations 2007 concern the management of health and safety. These regulations impose duties on clients, CDM co-ordinators, designers and contractors.

Consortium   The group of private sector participants who have come together for the purpose of tendering for a private finance initiative (PFI) contract. Also becomes a Special Purpose Company (SPC) or Special Purpose Vehicle (SPV). The generic term is the project company, which is established by the preferred tenderer and is the contracting party for a project (see Fig. 10.6, page 125).

Consultants   The client's or contractor's advisors on design, cost and other matters. Such advisors may include project managers, architects, engineers, quantity surveyors, accountants, bankers or other persons having expert knowledge of specific areas.

Contingency period   A period of time allocated in the tender works programme for contractor's time-risk events.

Contingency sum   An undefined provisional sum of money required by the employer to be included in the tender sum for unknown work.

Cost   The estimated cost of the physical production of work. Note that estimated cost should not be confused with historical cost; historical cost is the cost of construction which is revealed only after the work has been executed.

Cost records   Records of historical costs and notes of the conditions prevailing when such costs were incurred.

Domestic sub-contractors   Sub-contractors selected and employed by a contractor.

Down time (or Standing time)   The period of time that plant is not operating. This may be due to breakdown, servicing time or an inability to operate due to other factors.

Effective rate   The rate calculated by dividing a gang cost by the number of productive operatives in the gang.

Employer   The building owner or employing organisation, also called the client. The term employer is commonly used in UK construction contracts.

Estimate   Net estimated cost of carrying out the works for submission to management at the settlement meeting.

Estimating   The technical process of predicting the costs of construction.

Estimator   A person performing the estimating function in a construction organisation. Such a person may be a specialist or may carry out the estimating function in conjunction with other functions, such as quantity surveying, buying, planning or general management.

Firm price contract   A contract where the price is not subject to fluctuations during the contract period.

Fixed price contract   A contract where the price is agreed and fixed before construction starts.

Fluctuations   The increase or decrease in cost of labour, plant, materials and/or overhead costs that may occur during a contract.

Gang cost   A grouping of labour costs to include principal and supporting labour associated with a particular trade. It may also include items of plant.

General plant   Part of a contractor's project overhead calculation for plant, excluded from unit rate calculations, and which is available as a general facility on site. Durations for general plant are usually taken from the tender works programme.

Head office overheads   The incidental costs of running a business as a whole. They are 'indirect costs' that cannot be directly allocated to production, as opposed to 'direct costs' which are the costs of production. Among other things, overheads may include rent, rates, directors' salaries, pension fund contributions and auditors' fees. In accountancy, head office overheads are generally referred to as 'administrative expenses' whereas the direct costs of production are referred to as 'costs of sales'. The apportionment of head office overheads to individual projects or as a percentage of company turnover is decided by management as part of management policy.

JCT   The Joint Contracts Tribunal Ltd. A company comprising representatives of nine constituent bodies responsible for producing the standard forms of building contract.

Labour-only sub-contractors   Sub-contractors whose services are limited to the provision of labour.

Lump sum contract   A fixed price contract where contractors undertake to be responsible for executing the complete contract work for a stated total sum of money.

Management   Those responsible for the function of general management and having the responsibility for making the decision to tender and for reviewing tenders.

Margin   The sum that is required by an organisation, from a project, as a contribution towards its head office overheads and profit.

Market testing   A procedure for re-pricing the provision of services on a periodic basis by means of a competitive tender.

Mark-up   The sum added to a cost estimate, following the settlement meeting, to arrive at a tender sum. Mark-up will include margin, allowances for exceptional risks, and adjustments for commercial matters such as financing charges, cashflow, opportunities (scope) and competition. There may be a requirement for the main contractor's discount when tendering as a sub-contractor, and value added tax when required in the tender instructions.

Master programme   The name given to the contractor's overall programme for the works under the JCT form of contract. The programme should contain no more than 2000 activities of durations no greater than 1.5 times the progress reporting period, or it will be too cumbersome to maintain. It should be in a critical path network and show, among other things, all contract requirements of dates for possession and completion, the planned durations sequence, the logic of the principal activities and the critical path (or paths) to every completion date.

Method statement   A statement of the construction methods and resources to be employed in executing construction work. This statement is normally closely linked to a tender works programme.

Nominated sub-contractor/supplier   A sub-contractor/supplier whose final selection and approval is made by the client or client's advisors. (See 'Prime cost'.)

Open competitive tendering   An impartial method of procurement whereby contractors are invited through advertisements to apply for tender documents. The number of tenderers is not usually limited and reputation and ability to execute the work satisfactorily are not always considered.

Output specification   The specification that sets out the requirements in nonprescriptive terms, so that the tenderers can determine how to provide the services.

PDM   Precedence diagram method. A method of constructing a logic network using nodes to represent the activities and connecting them by lines that show dependencies.

PFI   Private finance initiative. PFI was initially developed by the UK to provide financial support for public–private partnerships between the public and private sectors. It has now been adopted throughout the world as part of a wider programme for privatisation and deregulation driven by corporations and governments and international bodies.

Pre-construction   Time preceding the construction stage, typically involving feasibility, design and procurement.

Preliminaries   The costs of running a site as a whole (rather than any particular zone of the site or any particular activities). They are sometimes referred to as site

overheads (or in the USA as 'field costs'). Thus, it is essential that each case is inspected individually to determine which resources are affected by a delay or disruption irrespective of how the contractor has priced that resource.

Pre-qualification   The provision of information by a contractor as part of a pre-selection process. An application by a contractor to be included on a select list of tenderers.

Pre-selection   The establishment of a list of contractors with suitable experience, resources, ability and desire to execute a project in relation to the character, size, location and timing of a project.

Prime cost (PC)   When used in bills of quantities and specifications, PC means that part of an estimate for work or supply of materials to be provided by nominated sub-contractors or suppliers. PC sums are determined by the client's advisors and detailed in the tender documents. The contractor may also be invited to execute work covered by a PC sum in certain instances. The sums inserted in bills of quantities are net of overheads, profit and attendances.

Project overheads   (sometimes referred to as site overheads, preliminaries, general cost items or expenses). The cost of those site-specific project costs that cannot be allocated to individual activities and which are not included in all-in or composite rates. Among other things, these costs may include site management, huts, safety precautions, job-related insurances, bonding costs, telephone, water, electricity costs, etc. The essential characteristic is that these overheads serve more than one activity, e.g. tower crane, skips, general site labour, etc. However, in practice some resources that could be allocated to an activity, e.g. scaffolding for falsework, are included in the preliminaries because of the contractor's preferred method of pricing.

Provisional sums   The Standard Method of Measurement for Building Works (SMM) provides for sums that may be included in tender documents for work which cannot be measured at the tender stage. These sums are inclusive of overheads and profit allowances. There are two types of provisional sums: defined and undefined. For defined provisional sums, contractors must be given full information about the nature and extent of the work and the contractor is then required to make provision in its tender works programme for an adequate duration and sequence for the work together with the associated preliminaries. Undefined provisional sums are for work which cannot be foreseen and, unless specifically indicated in the tender documents, the contractor is not required to include any duration for the work in its tender works programme nor any allowance for preliminaries. Work executed by reference to an undefined provisional sum is subject to a valuation. Where a JCT standard form of contract provides required instructions for the expenditure of an undefined provisional sum, such instructions constitute an event which may give rise to an extension of time and/or compensation.

Public sector comparator (PSC)   The PSC is an assessment of the scheme which includes capital costs, operating costs and third party revenues. The PSC is a benchmark against which value for money can be gauged. Clients use technical advisors to produce a reference project – sometimes called the public sector scheme (PSS).

Risk   Additional technical, contractual, financial and managerial responsibilities which form part of the contractor's formal obligations.

Scope   Opportunities to improve the financial, commercial or business aims of a construction organisation.

Selective tendering   A method of selecting tenderers and obtaining tenders whereby a limited number of contractors are invited to tender. The tender list is made up of contractors who are considered suitable and able to carry out the work. This suitability is usually determined by pre-selection procedures.

Service level specification   The specification given in the agreed project agreement setting out the standard to which the service must be provided. This is accompanied by an agreed performance monitoring regime.

Settlement   The action taken by management to convert an estimate into a tender. Also commonly known as 'appraisal' or 'final review'.

Settlement meeting   A timetable for the preparation of an estimate, all necessary supporting actions and for the subsequent conversion of the estimate into a tender.

Short-term programme   The name given to the contractor's strategic works programme for a particular section of the works. It should be in a critical path network, integrated with the master programme and show, among other things, the resources to be used on each activity and their duration and the cost calculated by reference to the productivity that the resource is planned to achieve.

Standing plant   Plant retained on site that is not working, but for which a contractor is still liable.

Temporary works   Resources needed for non-permanent work. Some temporary works such as formwork are measured in a bill of quantities, others such as hoardings are normally excluded from unit rate calculations because they are common to a number of activities and their durations are taken from the tender works programme.

Tender   A sum of money, time and other conditions required by a tenderer to complete the specified construction work. For design and build, the term tender includes design ('contractor's proposals') and price ('contract sum analysis').

Tender documents   Documents provided for the information of tenderers, in order to establish a common basis for their offers.

Tendering   The process of preparing and submitting for acceptance a conforming offer to carry out work for a price, thus converting the estimate to a bid.

Tender preparation programme   Resourced activity schedule outlining programme for preparation and submission of tender.

Tender settlement   The conversion of a cost estimate into a tender taking the commercial interests of the contractor into account. (See 'Settlement'.)

Tender works programme   A programme based on a global strategy for the project resulting from the information available at the tender stage. It should not be relied on for construction purposes but may legitimately be the basis for the construction master programme. Its purpose is to demonstrate that the contractor intends to comply with the date constraints listed in the tender documents and, where a contract period is not specified, it will indicate the completion date and hence, the duration over which the time-related site costs must be included in the tender. It should be prepared as a critical path network in order to identify critical work activities and delivery dates of client supplied information, goods and materials. It is not uncommon for the tender documents to require the tender works programme to be submitted with the tender.

TUPE   The Transfer of Undertakings (Protection of Employment) Regulations 1981.

Unavailability   The test for determining deductions from unitary payment by reference to standards for the provision of the facility for private finance initiative (PFI) projects.

Unitary payment   The payment by the awarding authority to the project company for the provision of the facility.

Variant bid   A bid which does not comply with the prescribed requirements of the awarding authority for a reference bid, but which a tenderer is proposing as offering better value for money.

Working Rule Agreement (WRA)   National Working Rules for the building industry produced by the National Joint Council for the Building Industry.

1 Introduction

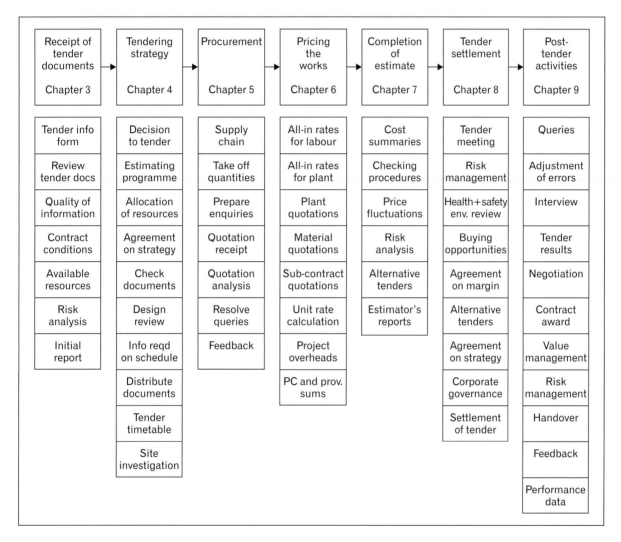

Receipt of tender documents Chapter 3	Tendering strategy Chapter 4	Procurement Chapter 5	Pricing the works Chapter 6	Completion of estimate Chapter 7	Tender settlement Chapter 8	Post-tender activities Chapter 9
Tender info form	Decision to tender	Supply chain	All-in rates for labour	Cost summaries	Tender meeting	Queries
Review tender docs	Estimating programme	Take off quantities	All-in rates for plant	Checking procedures	Risk management	Adjustment of errors
Quality of information	Allocation of resources	Prepare enquiries	Plant quotations	Price fluctuations	Health+safety env. review	Interview
Contract conditions	Agreement on strategy	Quotation receipt	Material quotations	Risk analysis	Buying opportunities	Tender results
Available resources	Check documents	Quotation analysis	Sub-contract quotations	Alternative tenders	Agreement on margin	Negotiation
Risk analysis	Design review	Resolve queries	Unit rate calculation	Estimator's reports	Alternative tenders	Contract award
Initial report	Info reqd on schedule	Feedback	Project overheads		Agreement on strategy	Value management
	Distribute documents		PC and prov. sums		Corporate governance	Risk management
	Tender timetable				Settlement of tender	Handover
	Site investigation					Feedback
						Performance data

Estimating is the systematic analytical calculation of projected construction and overhead costs for a contract. This Code of Practice is intended to describe the skills and set out the procedures that represent best practice from a commercial and professional point of view, as a practical guide for practitioners, students and academics. It should enable bid managers and chief estimators to establish and maintain sound and efficient systems, and for academics to introduce students to the underlying principles that they can take forward into their professional life.

The need for good estimating arises from the most common form of procurement in our industry: competitive lump sum tendering. Often, at some point in the supply chain, lump sum competitive tendering is required to enable the selection of contractors or sub-contractors to carry out contracts or sub-contracts. Despite the attempts by Sir John Egan and others to move the construction industry towards a less confrontational approach, competitive tendering still remains many clients' favoured route to arrive at best value. Methods such as management contracting and construction management move the major focus of competitive tendering 'down' the supply chain from principal contractor to package contractor selection. Private finance initiative (PFI) moves the first stage of competitive tendering 'up' the supply chain to the design, build and operate developer, and will usually require that estimators take much greater cognisance of whole-life costs, as opposed to purely capital construction cost. The growth of target cost/guaranteed maximum price contracts places additional demands on both the estimating and tender settlement functions.

Competition remains a fundamental requirement on most contracts for various links in the supply chain, making estimating a critical part of accurate and safe competitive tendering. When reading this Code of Practice, consideration should be given to the specific contracting approach that is being used, and where in the supply chain the estimating is being undertaken.

The construction industry has been under considerable pressure to innovate. Along with this has come a welcome change in procurement practice to move away from lowest price to best value as the leading selection criterion. The result of pre-construction practice has included an increase in demand for contractor input into the following:

- design and buildability

- value engineering

- project duration, sequence and timing

- demonstration of best value

- cost planning and 'design-to-cost' exercises including whole-life cost assessment.

Proposals for all of the above are often called for as a part of detailed tender submissions, giving clients the means to select contractors on the basis of much more than lowest price. This has changed the outputs of the bid process from the simple 'Form of Tender' to full-blown contractor's submissions incorporating design, planning and logistical proposals along with detailed costings. This consequently places far greater demand on a contractor's estimating process in terms of the quantity and quality of inputs required. In many respects, these changes leave the core of estimating practice unchanged, however, they alter the location and the significance of estimating in the business process. Many contractors have created multiskilled bid management departments incorporating their estimating departments, which previously took the lead in preparing tenders. The effect on the estimating process is to increase the number of inputs and outputs, which we show for a traditional approach in the flow chart at the start of Chapter 6. The flow charts for alternative contractual arrangements in Chapter 11, demonstrate the various demands of different forms of procurement, from the still popular and simple lump-sum tendering to PFI bidding.

Another aspect of seeking best value is an increase in the attention paid to the supply chain involved in tendering and delivering contracts. Specialist sub-contractors are seen as increasingly relevant in the demonstration of best value, as it is their resources that will be applied and which will be largely responsible for realising the benefits to be gained from innovation and technological development. The need for assurance of high-quality delivery from sub-contractors encourages main contractors to enter into formal and informal partnering arrangements with specialist contractors who can demonstrate a good track record. This changes the procurement approach for some contractors and some tenders, making the cost negotiations of work packages, with assurance of best value, a preferred route. The estimator's relationship with specialist contractors during the tender period must then become much more intimate, with more collaborative working on design development, methodology and cost estimating. The need to work in this way will also impose some restrictions on the way in which some estimators have, in the past, managed the process of procuring competitive sub-contract quotations.

The management of risk is a key thread that has been taken through the Code to emphasise the greater responsibilities being placed on contractors, along with the need to make sufficient allowance for health, safety and environmental issues.

In this context, this Code of Practice addresses the major problems facing estimators in preparing estimates which will form a substantial element of the more complex tender submissions now being demanded by clients. In addition, the Code will also represent a sound basis on which contractors can base their budgeting and post-tender cost control processes.

2 Underlying principles

2.1 Principles

It is important to recognise that the processes of estimating and tendering are only a part of the wider process of commercial management within a construction business. Whilst lowest price remains a key element for selecting contractors, more and more clients, particularly those in the public sector, are selecting contractors with an evaluation process based on quality and value addition, as well as commercial competitiveness.

In addition, it should be remembered that the tender is not just a means to secure a contract award. If the bid is successful, the estimate then becomes the basis for the contractor's construction budget and subsequent post-tender cost management strategy. It is therefore important to recognise and take account of the additional external pressures (for example, competitiveness) and internal constraints (for example, realistic recovery of projected preliminaries) that any estimator has to balance.

Good practice procedures are based on sound principles and, in this section of the Code, some of the underlying principles for estimating and bid management are set out. The following sub-sections are intended to give a brief introduction to these areas and indicate their relevance to this Code of Practice:

- general good practice
- bid management
- risk management
- value management
- health, safety, welfare and environmental issues
- supply chain management.

General good practice

The client's needs
When preparing a tender, a contractor is normally responding to an invitation to tender from a client. The client normally accepts the tender that is perceived to be the best to meet their needs. Establishing what the client wants from a project should therefore be of prime importance, and references back to these needs should be made in the estimator's report (see Chapter 7, Section 6). The tender submission document that accompanies a bid should demonstrate how these needs have been addressed.

Professionalism
Good practice involves acting in a professional manner in all dealings with clients and consultants, sub-contractors and suppliers. Such dealings should be honest, straightforward, thorough, accurate and informed. Communications should be

5

capable of being understood, appreciated and acted on. Practice professionalism adds authority to what is done, makes activities more effective and successful. It is essential to remember this point throughout the estimating process.

Investigation

The tendering phase of a project is the time when a number of thorough investigations, depending upon the nature of the tender, are potentially required. Chapters 3 and 4 explain the project appreciation process in detail. Where a design is provided with tender documentation, a global design review is essential in order to arrive at a clear understanding of what is required. A contractor should start a query list at the outset and seek to have meetings with the design team to get their explanation of the project and responses to queries (should the nature of the tender and the timescales permit). By being aware of the options available, a contractor can seek a competitive edge with better value (alternative) products or innovative techniques. Information gained during the tender stage can be essential to winning the tender, and ultimately to the successful completion of the project.

Appropriate procedures

This Code of Practice sets out procedures that can assist estimators and other bid preparation professionals to establish and maintain a set of procedures for estimating, tailored to the needs of their particular company. The procedures set out here may be unnecessary or too complex for some organisations. However, the underlying principles are generally applicable and should be borne in mind, and be reflected in the methods adopted.

Error avoidance

One of the basic principles of estimating is error avoidance; estimates cannot be guaranteed to be error free, but good procedures will keep errors and their impacts to the minimum. If a tender is unnaturally low due to errors it may be disregarded by the client, or alternatively, it may be accepted and the contractor could make a substantial loss. Clearly set out procedures are also essential when working as part of a team as they enable others to undertake work, to complete it or take it on to the next stage. Logical, transparent procedures enable estimators to explain and demonstrate what is required to new colleagues. When there is a successful tender, the construction team would expect the estimator to hand over material and workings that are clearly recorded, and the information used must be readily available with all accompanying forms, analyses and reports.

Leadership

Traditionally, estimators have led the bidding team at the tender stage, setting the agenda and facilitating the process. With the increasing demands for more sophisticated and detailed tender submissions, there has been the increasing use of a bid manager. Tender requirements have become much broader and now often require a multiskilled team to carry out the bidding process. On larger tenders, bid managers or even project managers will lead the tender team. On small or simple tenders, the estimator may still be required to lead the process, requiring skills in communication, team working and co-ordination. Estimators should see it as part of their professional duty to be able to take on a leadership role and indeed to progress to becoming bid managers.

Company policies

A tender is a particular company's solution to the needs of a particular project and, as such, must present that company's policies in all elements of the tender, from the method statements in the tender submission to the preliminaries, overheads and profit. If a tender does not comply with company policies, the construction team may not be able to complete the contract successfully. Estimators must keep themselves and their supply chain informed of the relevant aspects of company policies to make sure that they are incorporated into the tender costs. It is equally important to make the client aware of policies where they are relevant and where they meet a client's needs, and could add value to the tender. An estimator should ensure that the pricing of the preliminaries includes all necessary costs arising from the implementation of company policies, for instance on health, safety, welfare and environmental issues, and be able to demonstrate how the contract could help to meet the objectives of the company's business plan in terms of work-load and profitability.

Contractor selection

The increasing sophistication of construction clients and the complexity of their needs have led most of them to base their selection criteria for contractors on best value principles rather than lowest cost. Issues such as the proposed team, methodology, project sequence and duration, and value addition can be more important than cost in selecting the winning tender. Clients sometimes use selection panels, interviews and scoring matrices to assist in contractor selection. Where a collaborative form of procurement is used, it is likely that significant emphasis will be placed on an assessment of the extent to which the contractor's business ethics and core management philosophies match those of the client. Presentation of the tender submission documents, the individual team members and the company has therefore become an increasingly important issue, not just at the tender submission stage but also in mid-tender meetings and post-tender interviews. In order to demonstrate that a particular tender provides the best value it must adequately express the crucial issues in the proposal documents in written and drawn method statements, diagrams and design proposals. At a post-tender interview, it can be beneficial to utilise the latest technology alongside good presentation skills to communicate a proposal. Clients and consultants are usually well practised in interpreting tender submissions and scoring tender interviews. They will not be impressed by volumes of standard information taken off the shelf and reproduced in a tender submission simply to fill up space. More impressive are specific proposals that meet the needs of that particular project. Slick interview teams who will never play any further part in their project are unlikely to make a good impression; the team should be comprised of the people who will be working on the contract, and they need to demonstrate how they will approach the key project challenges.

The competitive edge

The cost of tendering rises in proportion to the increases in the requirements of tender submissions and it becomes all the more important for a contractor to maintain an adequate success rate when tendering. The estimator should aim to achieve the highest winning bid while maximising profit levels. This can only be done when the lowest, accurate estimated cost for the project is known. Effective procurement of sub-contract works is vital to this, as on a main contract tender they will often make up over 80% of the estimated costs. The estimator should also look for a competitive advantage, be it through alternative more economical suppliers, design simplifications (although this may incur design liabilities), or more cost-effective methods. Through careful interpretation of the client's requirements, as stated in the tender documents, an estimator can submit a very competitive tender with alternatives for the client's consideration.

Qualifications, or tender notes, can be used to set aside items that the estimator believes should be excluded and to limit risks, for instance by giving provisional sums for unclear items rather than adding a risk allowance to the tender (should the tender document allow). Such strategies, if clearly explained, are legitimate and, indeed, often necessary in competitive tendering. The risk and implications of a non-compliant tender should also be considered at the tender settlement stage by senior management (see Chapter 8).

Bid management

The estimator's role must be seen in the context of the overall bid process including:

■ sales and marketing

■ estimating

■ project planning

■ design management

■ procurement.

When the majority of construction work was carried out on the basis of traditional lump sum (competitive tendering without any design input), the tendering process was simple as the major task was estimating. In recent years, other forms of procurement including the many variations of design and build (D&B), have become more prevalent. For a large D&B project the bidding team may include:

■ architects

■ structural engineers

■ services engineers

in addition to the contractor's own:

■ planning engineer

■ project scheduler

■ construction manager

■ design manager

■ estimator

■ supply chain manager.

The tendering process under D&B is more complex and takes more time and resources than the traditional approach, making good management of the process essential. This is covered in more detail in Chapter 11.

Risk management

Risk management is increasingly seen as not only good management practice but also critical to the survival of any business. This Code is not intended to give detailed guidance on how to undertake risk management. Reference should be made to other specific risk management publications such as Risk Analysis and Management for Projects (RAMP) and Project Risk Analysis and Management (PRAM). Here, we are looking to identify the risks associated with estimating, and explain why risk management needs to be undertaken during the estimating stage of a project. The two most critical stages of tendering are the decision to tender and tender settlement. The risk management process enables contractors to identify, quantify and manage risks, thereby making better decisions and achieving better outcomes. This will be covered in Chapters 3 and 8.

Every organisation manages its risk, but not always in a way that is visible, repeatable and consistent in how it supports decision-making. The task of the Office of Government Commerce's (OGC) Management of Risk (MoR) is to enable any organisation to make cost-effective use of a risk process that incorporates a series of well-defined steps. The aim is to support better decision-making through a good understanding of risks and their potential impacts.

MoR provides a generic framework for the management of risk across all parts of an organisation: strategic, programme, project and operational. It incorporates all the activities required to identify and control the exposure to any type of risk, positive or negative, which may have an impact on the achievement of your organisation's business objectives.

The MoR approach was designed to complement the OGC's guidance on programme, project and service management (MSP, PRINCE2™ and ITIL) but covers the risk management elements in greater detail. This publication draws on experience from a variety of experts from the public and private sectors. It sets out a framework for making informed decisions about risk at a project, programme and strategic level to ensure that key risks are identified and assessed and that action is taken to address them.

Value management

This process runs alongside the design and procurement process as a dynamic cost review. It is of far more use when estimating for a D&B tender or indeed for a two-stage tender as a demonstration of value for money, may be a top priority for the client.

Value management gives the client assurance that he is getting the best value for his project, while maintaining his design, quality, programme and cost objectives. Advice on the value management process and the involvement of the contractor is included in Chapters 9 and 11.

Health, safety, welfare and environment issues

There are rapidly changing legal and moral obligations within the construction industry that need to be fulfilled with regards to health, safety, welfare and environmental (HSW&E) issues. There is a body of law on health and safety and corporate responsibility that must be adhered to. Operatives' welfare similarly is covered by law, but is also a practical employment issue since poor welfare facilities discourage and demotivate operatives and sub-contractors and leads to lower productivity. Whether we are obliged by law or policy, the cost of appropriate HSW&E issues must be addressed when preparing an estimate.

In the first instance, we must identify the key HSW&E issues, hazards and challenges including those imposed by CDM 2007 regulations. This is covered in Chapter 3 where an initial risk assessment of hazardous operations and conditions is outlined. Many of the issues will also be revisited during procurement, pricing the works and at the tender settlement stage.

By managing HSW&E issues effectively at tender stage, it may be possible to plan-out many problems by planning-in solutions. Demonstration of a company's approach to HSW&E on a particular project should add to the value of a tender and therefore may improve that company's tender success rate.

Supply chain management

The estimator is a crucial link in the supply chain and in supply chain management. For clients and consultants the estimator could be introducing specialist sub-contractors and suppliers who will be critical to the project success. Larger contractors will have a buying department or supply chain manager who may take the prime role in developing and maintaining relationships with suppliers and sub-contractors.

Most contractors keep an approved list of specialists for use in procurement. This will record information on performance on site and when tendering, financial and insurance details and HSW&E compliance. Such lists are critical to effective maintenance of performance quality and compliance to legal and policy criteria. A range of specialists is important as some projects require cost-effective and quick solutions, whilst others require innovation or high quality thinking. Keeping this type of information helps in using the appropriate specialist on a specific project. This will be covered in detail in Chapter 5.

Effective supply chain management that can be demonstrated to clients is an excellent contribution to the added value that a contractor can bring to a contract. It will give clients an assurance of the performance that a contractor will provide, particularly where they have not worked together before.

2.2 Processes

Project planning

All projects need logistics planning. Logistical solutions need to be found which may be simple or complex, depending on the project, and may be met by the contractor's own resources or by outsourcing. The project requirements and alternative solutions must be considered early on in the tender process and a strategy agreed in order to plan the project duration, sequence and timing, and the cost the necessary resources. This will be covered in more detail in Chapter 4.

Computerisation

Computers are tools which, if used properly, increase efficiency and improve the quality of work produced. Computers give estimators the opportunity to use experience and analytical skills to the maximum by eliminating the more mundane and low-skill arithmetic tasks. Information technology provides the opportunities to not only increase the accuracy and speed of estimating output, but also to carry out new and sophisticated tasks such as financial and time modelling, thereby improving the quality of the tender submissions made to clients. Chapter 12 looks at computerisation and sets out the best practice requirements for e-tendering; this is intended as a brief overview of the use of computers generally within the estimating process.

Pre-qualification process

Pre-qualification is concerned with the establishment by the client of a list of contractors or specialist trade contractors with the necessary skills, experience, resources, previous tender performance and desire to carry out the works, bearing in mind the character, size, location and timing of the project.

Pre-qualification will occur some time before any formal invitations to tender are issued. It is important for the client to allow a realistic programme covering the whole of the pre-qualification and tendering period, and that adequate time is given for each stage.

The contractor must be given adequate time to evaluate the material provided and prepare the information and data required to compile his tender. Comprehensive and detailed information will be required, and on major works, such submissions can take several weeks to prepare. The European Union (EU) has specific requirements regarding time periods when tendering for public sector works.

Contractor selection may be made by competition or negotiation, and is often by a combination of both. Pre-qualification procedures are usually used to draw up a tender list in order to invite competitive tenders.

Selection processes

Negotiation

Under specific circumstances, a contractor may be approached to negotiate a contract price without the need to introduce competitive tenders. In such cases, a contractor may be selected on the basis of past performance, recommendation, familiarity with the work or most commonly because of a close working or business relationship with the employer or consultants. This is admittedly more prevailing within the private sector.

Negotiation allows early contractor selection, especially where the design would benefit from the constructor's input. This can reduce the overall project programme, increase buildability and tailor the costs to the employer's budget. The counterargument is that the initial price may be higher and difficult to compare with competitive market rates. There are also procedures adopted by many (usually public) organisations to ensure that goods and services are procured by competitive means, particularly where finances are publicly accountable.

Open competitive tendering

This is a method of tendering which excludes pre-qualification and thus permits any applicants to join a lengthy tender list which can, and often is, beyond a sensible level. This arrangement is an approved EU option but it is expensive for clients.

Approved lists and framework agreements

This form of selective tendering enables clients to choose tenderers for a project from a standing list of contractors, who have been vetted and pre-qualified for various categories of work at an earlier stage. Contractors are asked to apply for categories, which are defined by contract value and nature of work. In a framework agreement there may be some guarantee of the number of enquiries to be sent to a contractor in a given period.

It is important that clients and clients' representatives monitor and regularly update their lists of contractors to:

■ exclude companies whose performance has been unsatisfactory

■ introduce suitable new companies that can demonstrate the required qualities and abilities

■ compile the lists in a form appropriate to the class of project

■ include companies with the financial capacity and stability to carry out the work.

One-off project (*ad hoc*) lists

Clients, or their consultants, frequently create an initial list of suitable contractors solely for a particular project. There will often be a pool of contractors from which to choose that can be assembled in three ways:

■ by including contractors who write with an early expression of interest in a scheme

■ by using an advertisement to invite applications, or

■ clients or consultants may draw on their experience or interrogate their database.

Contractors included on an initial list are usually asked to provide information about their financial and technical performance, particularly in relation to the type of work under consideration. More elaborate pre-qualification practices include completion of questionnaires and making presentations to the clients and their consultants. An assessment of a contractor's competence in compliance with health and safety legislation is now a statutory requirement prior to the award of a contract, and so may be dealt with at the pre-qualification stage. The EU procurement rules are more prescriptive as to what information should be provided by contractors about themselves.

3   Receipt of tender documents

Initially, the contractor will be concerned with a quick overview of the tender required for a project. This will enable the contractor to plan the workload to be undertaken by their estimating department, after having checked that the project suits their particular skills and meets business objectives. The contractor should also ensure that adequate management and resources are available to support the project during the tendering period. A risk analysis would be undertaken and an initial report would then be submitted to the contractor's management team.

A contractor should formulate objectives and a strategy to win contracts at prices that will produce a profitable outcome. A bidding strategy must also recognise the need to decline invitations to tender for work which fail to meet the organisation's objectives.

3.1   Tender information form

Receipt of tender documents

On receipt of tender documents a tender information form (see Fig. 3.1) should be completed by the estimating department and an acknowledgement or receipt of the enquiry sent to the client or the appropriate consultant. This form will provide management with a summary of the project and the tender documentation; it will be circulated to managers in appropriate departments, and will be significant in contributing to the decision for the submission of a tender.

Figure 3.1 A specimen tender information form.

	TENDER INFORMATION FORM	Tender number

PROJECT PARTICULARS

Project title	Client
Project address	Architect
Project description	Engineer
Contact for site visit:	Quantity surveyor
Drawings available at:	

TENDER INFORMATION

		Estimated cost breakdown:	£
Date enquiry received		Own work	
Date for tender submission		Domestic sub-contractors	
Tender validity period		Project overheads	
Type of tender open ☐ negotiated ☐		PC & provisional sums	
selective ☐ other ☐			
Documents BOQ ☐ no BOQ ☐		Total	

CONTRACT DETAILS

Form of contract	Method of measurement	
	Period of interim certificates	
Amendments	Payment intervals	
	Period for honouring certificates	
Deed	Retention	
	Defects liability period	
Insurances	Liquidated damages	
Bonds	Fixed price ☐ Fluctuating price ☐	
Warranties	Programme Start	
	Duration	

Notes

To make a realistic judgement on whether to tender for a project, a contractor should ideally be in possession of the following preliminary information.

Client

- Details of the client, or if a subsidiary company, details of the holding company.

- Full particulars of consultants to be used on the project, including their duties and responsibilities.

- A description of the tender documents, their expected date of issue, the period available for tendering, the acceptance period for the tender and time when unsuccessful tenderers will be notified.

- Whether the project, either in its present or a different form, has been the subject of a previous invitation to tender (if available).

- The latest date for receipt of acceptance of invitation to tender.

- The number of tenders to be invited (if available).

Project details

- The location of the proposed works, including preliminary drawings and a site plan.

- Description of the project.

- Approximate cost of the project.

- The intended date for commencement of the contract.

- The intended period for completion, if any.

- Details of any phasing and/or sectional completion.

- An outline of the form of construction.

- Access issues.

- Special operational space requirements.

- Ground conditions.

- Sufficient dimensions and specification details to permit evaluation of the project.

- Details of work to be carried out by nominated sub-contractors, approximate value, and names if known.

- An indication of health and safety issues in form of the Pre-Construction Information (from CDM Coordinator).

The decision to tender can be made at one of two stages:

- when pre-qualification enquiries are initiated by a client or his consultants, the contractor will make a decision based on an outline of the tender information available at that stage. This intention to submit a tender must be reaffirmed when the full invitation to tender and supporting documentation are received

- when the pre-qualification procedure has not been followed the contractor may find that invitations to tender arrive without prior notice. In such instances, only one opportunity exists to appraise the project and make the necessary decision to tender for a project or not.

Projects for which pre-qualification has not occurred

When the contractor receives tender documents for a project there is increased urgency in deciding whether to tender (or not). The documents may be unexpected (although this is usually not the case) and cause problems in an estimating department which already has a full workload. The project may be of particular interest to the contractor and so it is vital that an early appraisal is made of the potential of any new enquiry.

If the project cannot be accommodated because of workload in the estimating department, or the company's workload needs, then the client must be advised, at the earliest opportunity, that a tender will not be submitted. This will allow time for the selection of another contractor by the client or consultant, if it is considered necessary to retain a full tender list.

If the initial assessment of the tender documents indicates that the project is of interest to the contractor, and if estimating department workload commitments allow, then a full inspection of the tender documents will be necessary.

Projects where pre-qualification has occurred

The first operation will be to check that the tender documentation conforms with the information given at the time of pre-qualification. If this initial inspection discloses discrepancies in the project information, programme, conditions of contract or other areas, the tendering authority should be advised accordingly. The contractor will then need to examine the tender documents in detail to establish whether the changes are of such significance as to discourage tendering.

If the initial inspection confirms that the project conforms to the information given at the time of pre-qualification, the contractor can then proceed with a detailed examination of the tender documents.

| 3.2 | **Review tender documents** |

An inspection of the tender documents must be made by the estimator responsible for the production of the cost estimate and a checklist must be established of the documents received. In larger organisations, the documents may be inspected by the chief estimator and other members of the contractor's organisation, including the planning engineer, project scheduler, quantity surveyor, buyer, contracts advisor and contracts manager.

Clear lines of communication are needed to ensure that all viewpoints of those examining the documents are considered. The estimator (or for larger projects, the bid manager), will be responsible for the co-ordination of these views.

Inspection of the tender documents must seek to achieve, as a minimum, the following objectives:

■ the documents received are those for the project under review

■ the documents and information are adequate for assessing costs

■ sufficient time is available for production of the tender.

Where an approximate cost of the project is provided this must be reconciled against any advice given at the time of pre-qualification. If no approximate cost is given, an early assessment must be made by the estimator to determine the approximate cost of the project and the scope of the works.

3.3   Quality of information

The estimator is particularly concerned with the quality of the information provided at the enquiry stage. Inadequate or contradictory information leads to a cautious approach by the estimator. While the estimator should seek further information from the tendering authority, in such cases, the estimator's approach is often influenced when uncertainties exist and the project is not fully defined. Past experience may well indicate that the uncertainties continue into the construction phase with negative consequences, for both the project and the contractor.

Normally the project specification is produced in a standard format clearly linked to the Common Arrangement of Works Sections for Buildings Works (CAWS). Where a bill of quantities is provided, any deviation from the Standard Method of Measurement must be noted. Where bills of quantities are not provided, good drawings together with a comprehensive specification will be required depending upon the procurement route chosen.

The quality of design and evidence of co-ordination of services will be of significance to the estimator. The extent of design advancement, as reflected in the drawings provided, must be compared with the bills of quantities to ensure that they accurately reflect the designed work. Note must be taken of provisional and prime cost (PC) sums that indicate where design is not yet complete.

The estimator must obtain from the project information – usually contained in the preliminaries clauses, in conjunction with the Pre-Construction Information – details which:

- affect the contractor's intended method of working
- impose restrictions
- affect access to the site
- interrupt the regular flow of trades
- affect the duration of the project
- affect the sequencing of the project
- require specialist skills or materials
- have a significant effect on the project sequence
- are of major cost significance.

Such items influence costs and will guide the estimator during the production of the estimate.

The estimator (or bid manager) must co-ordinate the views of the various members of the contractor's team who examine the project information and must provide management with a realistic appraisal of this information to enable a decision to be made on whether to tender for the project.

3.4   **Contract conditions**

The terms and conditions of contract often have an impact on any decision to tender and the following points should be considered and included in any risk analysis:

- form of contract to be used

- provisions for liquidated damages

- proposed amendments to standard forms of contract and appendix

- details of interim payments

- retention conditions

- bonding requirements

- details of insurance

- provisions for fluctuations in cost

- warranties and insurance policies required.

Many standard forms of contract will be encountered in building. These include:

- JCT 2005, with or without quantities alternatives, two stage, design and build and management contracting alternatives

- Intermediate Form tailored for medium-sized building projects, needing a less complex contract

- NEC3 suite of contracts used on many partnering projects

- the JCT Agreement for Minor Building Works used on smaller projects.

Older editions of standard forms and other more specialist forms are used by larger public sector and private companies. These could include PFI contracts, IChemE Red or Green Books and even the FIDIC forms. The risk balance with all these contracts varies and should be analysed at the project appreciation stage.

The particular contract conditions must be identified and any changes to standard forms of contract noted. Minor alterations can have a significant effect on the contractor's cash flow, the funding needs of the project, the risk allocation and the responsibilities of the respective parties to the contract. Such changes must be highlighted by the estimator (or contracts specialist) and the information passed to management on the extent of the inherent risk to the contractor of any changes made. Such advice will be of significance when deciding to tender.

Typically, such amendments may include time-bars on entitlement. For example, clauses rendering notice of delay within a prescribed period following an event, and the entitlement to an extension of time or compensation, for what would otherwise be at the employer's risk. Others may require notice that an instructed variation will have a delaying effect that is likely to adversely affect completion, and this risk is passes to the employer.

Supplementary conditions are often attached to contracts dealing with insurances, warranties, design responsibilities and performance bonds. Typically, contract bills and specifications may require the contractor to carry out design work where the contract conditions contain no design provisions in anticipate that the design will be prepared by the employer. Where onerous non-standard conditions are attached to a contract, many contractors either ask to be excused from tendering or add a premium to the tender bid.

3.5   Available resources

A decision to tender must also consider the resources currently available to the company.

Two separate decisions must be made concerning the workload. The chief estimator must make the decision based upon the workload of the estimating department, and be satisfied that the estimator who will be allocated to a project has the necessary expertise and knowledge needed for that particular task. The advantages of pre-planning, which will be possible from advance warning by pre-qualification, will be obvious.

The second decision must be made by company management. This concerns the objectives and needs of the company in terms of current and estimated future workload and also the availability of resources to construct the project. Management must be satisfied that the project meets the company's objectives regarding workload and that the company is not exposed to undue risk by committing to excessive levels of work with a single client, one project or one particular sector of industry.

3.6   Risk analysis

Risk analysis is a process that will take place throughout the preparation of the tender. Here we look at what risks need to be analysed before a decision to tender can be made. Should the project appear to attract an unacceptably high level of risk then no further time should be spent on preparing the tender and it should be declined.

Risks can be separated into those to be borne by the employer and those to be borne by the contractor. In each category, they can then be classified as known risks, known unknowns (uncertainties) and unknown unknowns (*force majeure*). The principal risks are identified that typically might fall into the following categories:

- health and safety
- environmental
- design
- construction
- project planning and scheduling
- innovation
- existing conditions
- availability of resources
- external factors
- client
- consultants
- terms and conditions of contract.

These risks are then assessed for their potential severity, i.e. the impact if they were to occur, and probability of occurrence, the product of which gives a score that leads to the assignment of a high, average or low risk rating. Alternatively, they could be given a financial allocation so that the cost effect of a strategy is known. It is important to manage all those identified as contractors' risks which attract a high or average risk rating by implementing an appropriate risk management strategy. Typical strategies include:

■ risk avoidance, such as a change in design which avoids a hazard

■ risk transfer, where a risk is passed on to another party such as a sub-contractor who will be better able to manage it

■ risk mitigation, where measures are taken to reduce the impact of a risk to the project

■ insurance against a risk (a form of risk transfer)

■ planning the activities likely to be affected as non-critical if possible

■ making cost and time contingency allowances to cover any residual impact of a risk.

3.7 Initial report

An initial report can now be produced to enable the management team to make the decision whether or not to accept the invitation to tender. A summary of the findings from the points discussed in this chapter would form the content of this report.

4 Tendering strategy

```
                    ┌─────────────┐
                    │  Tendering  │
                    │  strategy   │
                    │             │
                    │  Chapter 4  │
                    └─────────────┘
                    ┌─────────────┐
         Inputs     │  Decision   │     Outputs
                    │  to tender  │
                    ├─────────────┤
                    │ Settlement  │
                    │  meeting    │
                    ├─────────────┤
                    │   Tender    │
                    │ preparation │
                    │ programme   │
                    ├─────────────┤
                    │ Allocation  │
                    │of resources │
                    ├─────────────┤
                    │  Agreement  │
                    │ on strategy │
                    ├─────────────┤
                    │   Check     │
                    │ documents   │
                    ├─────────────┤
                    │ Information │
                    │  required   │
                    ├─────────────┤
                    │   Design    │
                    │   review    │
                    ├─────────────┤
                    │    Site     │
                    │investigation│
                    └─────────────┘
```

4.1 Decision to tender

The initial report produced following inspection of the tender documents will be taken into account by management in deciding whether a tender is to be submitted. In order to comply with company governance, the decision must be made by a director of the tendering company, taking into account the views of the other board members and the risk analysis carried out by the bid manager or chief estimator. The cost of preparing a tender must be viewed in relation to the chance of being awarded the contract.

Many firms might categorise the decision into one of the following groups:

■ priority tender (fully committed), or

■ regular tender.

Figure 4.1 A specimen tender register.

	TENDER REGISTER				start _____ finish _____

tender number	client	title	tender		result		notes
			date	sum	yes	no	
10165							
10166							
10167							
10168							
10169							
10170							
10171							
10172							
10173							
10174							
10175							
10176							
10177							
10178							
10179							
10180							
10181							
10182							
10183							
10184							
10185							
10186							
10187							
10188							
10189							
10190							
10191							
10192							
10193							
10194							
10195							
10196							
10197							
10198							
10199							
10200							

If it is decided to accept the invitation, the procedure for the preparation of an estimate described in Chapters 5 and 6 should be started and the tendering authority informed that a tender will be submitted. If it is decided that the invitation to tender is to be declined, a response should be given and all documents should then be returned immediately to the tendering authority with reasons for decline. This action, if taken, must be carried out as soon as possible after receipt of the tender documents.

In either case, the tender information form (Fig. 3.1) should be filed in a tender register (Fig. 4.1) with the decision noted and circulated to departmental managers.

4.2 Settlement meeting

The estimator (or the bid manager) must, as the person responsible for the production of the estimate, ensure that the timetable, which highlights the key dates in the production of the estimate and tender, is clearly annotated. This will be an essential document for all those associated with the tendering function. The following dates must be established without ambiguity:

- latest date for dispatch of enquiries for materials, plant and sub-contracted items

- latest date for the receipt of quotations

- bills of quantities production for D&B, drawings and specification contracts

- visit to the site and the locality

- finalisation of the method statement

- completion of pricing and measured rates

- finalisation of the tender works programme

- intermediate co-ordination meetings within the contractor's organisation

- review meetings

- submission of the tender.

All personnel associated with the tender must confirm that they are able to provide the necessary data in the format required, in accordance with the agreed timetable.

4.3 Tender preparation programme

The settlement meeting will be of significance to both the estimator and management. The time available for the production of the tender should be apportioned carefully to allow for the assimilation of project information, obtaining quotations from trade specialists and suppliers, and for completing the cost estimate.

In conjunction with the bid manager or chief estimator, the planning engineer and/or project scheduler should prepare a fully resourced critical path network of the estimating activities, indicating the activities, durations, sequence and timing necessary for the completion of and submission of the tender. Co-ordination meetings are needed with management and other departments within the contractor's organisation to establish key dates, decide on actions necessary, as well as monitor progress during the production of the cost estimate.

The tender preparation programme should be established for the benefit of the estimating department to show:

■ who is involved in preparing the estimate

■ what activities they will be doing

■ the dependencies between the activities

■ interim deliverables completion dates

■ activity durations based on man/day resource inputs

■ overall completion date

■ tender submission date.

The time available for preparing the estimate is usually short, and a tight control over activities is essential. Checklists and records are important to ensure no items have been overlooked.

4.4   Allocation of resources

Resources include finance, staff and labour (bearing in mind the particular skills and quality needed for a project), as well as the availability of materials and plant. The estimator contributes to this assessment with his knowledge of market conditions, trends in the costs of basic resources and in the identification of problem or risk areas.

The financial appraisal involves an assessment of the payment proposals, any bonding arrangements and the relative proportions of work to be carried out directly by the contractor, sublet to domestic sub-contractors or carried out by nominated sub-contractors. From these assessments, the funding requirements of the contract will be established and finance charges can be calculated.

Commonly, a contractor will provide his own operatives for materials handling, keeping the site clean, general building and attendances, and will outsource scaffolding, craneage and plant handling. On larger contracts a logistics contractor could be employed to provide a complete logistics package providing:

■ site security, access control, identity cards and surveillance cameras

■ materials storage, movement and management

■ stock control and planning

■ waste removal and recycling

■ transport, lifting, hoisting and craneage

■ welfare and office facilities, and associated supplies and cleaning.

4.5   Agreement on strategy

Many projects present unique problems and risks that must be identified, planned around and costed during the tender period. These may include:

■ restricted periods of possession or restricted working calendars (e.g. ground works)

■ confined inner-city sites with difficult access and limited storage space

- working alongside railways or airports with stringent safety and security requirements

- working adjacent to housing or occupied offices where noise restrictions will apply

- hazardous locations such as contaminated sites or over water

- fast-track programmes for essential facilities or emergency repairs.

The bid management team must identify and analyse all such issues relating to a project by reference to their own expertise, that of construction managers in their organisation, external advice from experts and specialist sub-contractors. Special solutions can thereby be adopted for challenging projects such as:

- just-in-time deliveries

- off-site storage and prefabrication

- adoption of stringent site rules and procedures covering safety and security

- pre-selection of sub-contractors experienced and qualified in railway or airport contracts

- special measures to limit and contain noise and other sources of nuisance

- site decontamination prior to commencement of general works

- access solutions that provide safe working platforms above or beside site hazards

- 24/7 working and triple shifting with special health, safety and welfare facilities and transportation, to make working conditions as acceptable as possible for operatives facing particularly demanding requirements

- restricted possession periods

- restricted working periods.

All such solutions must be fully researched and planned, including all necessary risk analyses, diagrams, drawings, flowcharts, critical path networks, work schedules and method statements, so that they can be costed and demonstrated to consultants and clients. The costs will be built into the tender preliminaries and/or works and the written and drawn information will form part of the tender submission. Any specific client's requirements for the project should be established as these may impact on the cost. Ensuring what would constitute a compliant bid needs to be clearly established. This could include:

- selection criteria (e.g. a high level of quality in the finished project may require high levels of supervision)

- method statement

- project duration, sequence or timing

- sectional completion and/or phasing requirements

- security conditions for sensitive sites, in relation to either the site or its surroundings

- provision of drawings of services as installed, for maintenance purposes

- commissioning and testing

Figure 4.2 A specimen tender documents checklist.

	TENDER REGISTER			Tender number			

	date received	number of copies	Distribution					
			planning	commercial	buying	design & build		
						architect	engineer	QS
FORMAL TENDER DOCUMENTS								
Invitation to tender								
Tender instructions								
Form of tender								
Return envelope								
DRAWINGS								
Architectural								
Engineering								
Services								
SCHEDULES								
SPECIFICATIONS								
Building								
Engineering								
Services								
BILLS OF QUANTITIES								
Building								
Engineering								
Services								
SCHEDULE OF WORKS								
SCHEDULE OF RATES								
Project specific								
General								
CONDITIONS OF CONTRACT								
Articles of agreement								
Appendix details								
Payment schedule								
Programme								
Design liabilities								
Parent Company Guarantee								
Bond								
Warranties								
TECHNICAL REPORTS								
Health and Safety Plan								
Method statements								
Site investigation								

■ training of client's staff in operating services or mechanical equipment

■ some advice or recommendations may also be sought.

This detailed plan can also be incorporated into a demonstration to the client on the value of employing a particular contractor. This planning exercise will give the client and his team the assurance they need that a difficult contract will be delivered successfully.

4.6 Check documents

Once the decision has been taken to tender for a project, the estimator must ensure that all the tender documents have been received. A check must be made to see that all drawings received are of the revision noted, and that all other documents listed in the invitation letter are provided. Figure 4.2 shows a checklist for tender documents which will help the estimator to assess their suitability for the estimate. A letter should be sent to the tendering authority to acknowledge the receipt of the documents and confirm that a tender will be produced by the due date. This letter should also record any discrepancies in the documents received.

Drawing lists can be checked against drawing issue sheets that usually accompany the documents. It is not necessary to produce issue sheets for suppliers and sub-contractors because their documents are recorded on enquiry abstracts and listed in enquiry letters. A record of any subsequent additional or revised documents should also be carefully recorded in the estimator's file.

It is important to make an accurate record of tender documents received because they will form the basis of a formal offer and eventually will be checked against the contract documents in the event of a successful tender. This can best be achieved by stamping each drawing and document to show the date received and the words 'tender document'. This procedure will later help to indicate their purpose and to differentiate them from construction issue-documents once a contract has been awarded. Where drawings are copied for sub-contractors and suppliers, they should similarly be stamped to avoid confusion between 'tender' and 'construction' copies.

Checking validity

In the case of a project where pre-qualification has occurred and details have already been sent to the contractor, a checking procedure is needed to confirm that the project conforms to the information already provided and that the contractor's position regarding tendering has not changed. It is essential that adequate time is allowed for this procedure by the tendering authority. In this situation, the contractor will follow the same procedures in analysing the information received.

4.7 Information required

For the preparation of a tender works programme, estimate and tender, all the information detailed above will be required, together with additional information. This is supplied in various forms, including:

■ drawings

■ specifications (including performance specifications where appropriate)

■ schedules

■ technical reports

- scheduled work periods, sequence and dependencies for critical specialist sub-contractors and nominated sub-contractors

- bills of quantities.

All documents should show the date of original issue and the date and nature of subsequent amendments. The drawings listed in the bills of quantities or drawing issue forms will be issued with the tender documents.

Site layout drawings

Drawings should show the existing circumstances of the site and adjoining areas and would usually include:

- site boundaries

- means of access, including approach roads and entrances, and any restrictions on access

- contours

- roadways and fences

- wooded areas, water and other natural features

- overhead, surface and underground lines and equipment of statutory undertakings and other owners

- position of buildings on site to be demolished

- position and height of buildings on site to be retained

- position and height of structures adjoining the site

- position and depth of existing foundations and groundworks

- temporary structures

- outline of new buildings and site works

- working areas

- working periods

- restricted areas

- location of strata investigations

- existing services on or near the site

- block plan showing roads, natural features and site orientation, sufficient to permit the location of the site to be ascertained.

Other drawings

Other drawings may include:

- general arrangements for the project, shown by plans, sections and elevations

- works to adjoining structure

- details of items requiring mechanical lifting, including size, weight, location and fixing needs

- in the case of alteration work, surveys of the existing structure

■ specifically designed structural temporary works

■ special risks, construction methods proposed and sequence related to engineering considerations.

Detailed drawings, outline drawings and schedules

Detailed drawings will be required to give sufficient information to determine the location, nature and dimensions of work to the following where appropriate:

■ site clearance and groundworks

■ sub-structure

■ frame

■ upper floors

■ roof

■ canopies, balconies and projections

■ staircases

■ lifts and other mechanical vertical circulation

■ external cladding

■ non-standard joinery

■ purpose-made doors and metal windows

■ structural walls

■ non-structural walls

■ suspended ceilings

■ finishes

■ internal drainage

■ external drainage

■ external work

■ builder's work in connection with services.

Outline drawings will be required to give information sufficient to determine the location and nature of work to:

■ waste, soil and overflow pipes

■ cold water services

■ hot water services

■ heating services

■ ventilation services

■ gas services

■ electrical services

■ communications and security services

■ other services.

Outline drawings or schedules may be required for work to:

- partitions
- fittings
- sanitary fittings
- non-standard joinery.

Schedules may be required (if not available elsewhere) for:

- standard joinery
- standard windows, frames and ironmongery
- standard doors, frames and ironmongery
- miscellaneous ironmongery
- external wall finishes
- internal wall finishes
- floor finishes
- ceiling finishes
- decorations.

Technical reports

- Site investigation report, including water table and an interpretation of factual data. This information is needed for piling, excavation methods, disposal of surplus excavated material and to consider the effect of chemicals on permanent works.

- Construction (Design and Management) Regulations (CDM) health and safety plan that identifies hazards likely to be encountered during construction, stating where and when they are likely to occur. The risk of a particular hazard occurring must be assessed by the CDM co-ordinator. Before tendering, the safety plan (generated from the Pre-Construction Information) must be sufficiently developed for it to form part of the contractors' tenders.

- Special site conditions such as fire risks, security risks or radiation hazards.

- Other technical reports, such as the condition of an existing structure or special requirements for temporary works or plant.

- Hazardous materials such as asbestos, contaminated ground and the need for decontamination of buildings.

4.8   Design review

Here the contractor should be reviewing the design (if one is provided) and dividing the project into suitable packages for various suppliers, sub-contractors and consultants as required by the particular project. An analysis of the gaps in provision should be made and queries to the design team submitted, if appropriate.

Visit to consultants

The estimator may need to visit the consultants, particularly when further information is needed which has not been given to tenderers, such as additional drawings and site investigation reports. In some circumstances, these documents may be confidential. Visits will normally be made to the architect, but it may also be advantageous to visit the consulting engineer, services engineer and quantity surveyor, in order to meet the personalities who will subsequently be involved with the project.

Detailed drawings, reports of site investigations and any other available information must be inspected and notes and sketches made of all matters affecting either construction method, temporary works or the likely cost of work.

A critical assessment must be made of the degree of advancement and quality of the design, for example with respect to RIBA design stages. A well-developed and well-documented design may be indicative of a smooth running and possibly profitable project. A design which is obviously deficient and incomplete may cause delay to progress, the time or cost effect of which may not always be at the risk of the employer. Clearly defined contingency allowances should be considered in the tender. An adverse report may lead to the reconsideration of the decision to submit a tender.

Visits to consultants have become less common because contractors are usually given copies of all the drawings and specifications which are going to become part of the contract documents. It is more likely that a visit will be seen as an opportunity to show that the tenderer is willing to contribute to the scheme, to work closely with the design team and to express an interest in further work.

Identifying missing information and establishing a list of queries for the consultants should be undertaken so that letters can be written during the tender period or raised at a mid-tender interview. The opportunity to review the answers to questions raised by other contractors is also essential as issues might affect the risk assessment or cost of the work.

4.9   Site investigation

A visit to the site should be made as soon as a preliminary assessment of the project has been carried out and a provisional method and sequence of construction established. The estimator may be accompanied on such a visit by other members of the tendering team.

As well as visiting the site itself, the opportunity should be used to examine the general locality and to establish the extent of other building works in the area. Visits should also be made to local labour agencies and suppliers in the area. Any opportunity to visit excavations near to the site must not be missed and in this connection, the local authorities may be able to give advice on local conditions and of any excavations which may be visible, adjacent or near to the site.

A comprehensive report of the site visit should be prepared and a sample form is shown in Fig. 4.3. Site photographs can often be a useful way of recording information for discussion and record purposes. A brief site visit video is particularly useful to present at the subsequent settlement meeting.

Points to be noted when making a site visit include:

■   position of the site in relation to road and rail and other public transport facilities

■   names and addresses of local and statutory authorities

■   topographical details of the site, including note of trees and site clearance required

■   any demolition work or temporary work needed to adjacent buildings

Figure 4.3 A specimen site visit report (continued on next page).

	SITE VISIT REPORT	Tender number
		PAGE 1 OF 2

PROJECT PARTICULARS	
Project title	**Contact for site visit**
	name
Project address	tel number
	fax number
	Visit details
Directions to site	person making visit
	date time

	REMARKS
SITE POSITION	
In relation to public transport	
Distance from local office	
Other work in the area	
Adjacent buildings	
Fencing and hoardings	
Demolition	
Hazards	
SITE CONDITIONS	
Topography	
Trees and vegetation	
Site clearance	
Ground conditions: borehole details	
type of soil	
stability of soil	
water table	
tidal conditions	
pumping requirements	
disposal of water	
Security problems	
Weather exposure	
Space for temporary accommodation	
Restraints for static plant: cranes	
scaffolding	
Live services	
Protection requirements	

Figure 4.3 A specimen site visit report (continued).

	SITE VISIT REPORT	Tender number
		PAGE 2 OF 2

	REMARKS
ACCESS	
Temporary roads	
Safety	
Deliveries	
Traffic restrictions	
LOCAL FACILITIES	
Disposal of soil	
Services: telephone	
water	
electricity	
sewers	
Garages, refreshments	
Local authorities	
Statutory authorities	
LOCAL CONTACTS	
Security services	
Labour agencies	
Plant hire	
Local tip and charges	
Sub-contractors	
Suppliers	
Sketches/photographs attached	

- access points to the site and any restraints on layouts that have been considered

- ground conditions, and any evidence of surface water or excavations indicating ground conditions and water table

- facilities in the area for the disposal of spoil

- existing services, water, sewers, electricity, overhead cables, etc.

- any security problems, evidence of vandalism, need for hoardings, lighting, etc.

- labour situation in the area

- availability of materials

- weather conditions: high rainfall, winds, etc.

- temporary roads

- location of nearest garages, hospital, police and cafés

- nature and use of adjacent buildings, such as industrial or residential

- police regulations

- local sub-contractors

- restraints imposed by adjacent buildings and services, i.e. space available for tower cranes, overhang, etc.

- other work currently in the area, or shortly to start

- availability of space for site offices, canteen, stores, toilets and storage

- the effect any client requirements may have on access, storage, movement or accommodation

- special difficulties.

HSW&E issues can be identified from reviewing the tender documents. However a site visit will be essential to see at first hand the conditions under which the contract will be carried out. Solutions can then be developed to address issues such as safe access, appropriate site facilities and environmental and considerate contractor measures required. The estimator will usually be assisted in this task by construction managers, or a health and safety manager.

The proposed solutions and methods will then either be included in the preliminaries or be passed to sub-contractors for inclusion in their quotations, or be used in order to prepare costings for specific items such as scaffolding, site accommodation and waste removal and recycling. Valuable advice can be obtained from specialist sub-contractors on alternative solutions and safe methods of working. Quotations received will form important parts of the preliminaries section of a lump sum tender including the site establishment, access solutions, craneage and lifting. Therefore, it is particularly important that these fully take into account HSW&E issues and company policies.

HSW&E issues must also be high on the agenda at tender review meetings. If a design is inherently unsafe or, the client is not prepared to make due allowance for adequate site welfare facilities, it may be necessary to decline to tender or insist that HSW&E issues are properly taken into account by the client and their consultants. When it comes to tender settlement, the company directors will need to see that their legal responsibilities, as well as company policies, have been taken into account in the proposed methodology, resources and costings.

5 Procurement

```
┌─────────────────────────────────────────────────────────┐
│                      ┌──────────────┐                     │
│                      │ Procurement  │                     │
│                      │              │                     │
│                      │  Chapter 5   │                     │
│                      └──────────────┘                     │
│                                                           │
│         Inputs       ┌──────────────┐     Outputs         │
│                      │    Supply     │                     │
│                      │    chain      │                     │
│                      ├──────────────┤                     │
│                      │   Take off    │                     │
│                      │  quantities   │                     │
│                      ├──────────────┤                     │
│                      │ Preparation   │                     │
│                      │     of        │                     │
│                      │   enquiry     │                     │
│                      │  documents    │                     │
│                      ├──────────────┤                     │
│                      │  Quotation    │                     │
│                      │   receipt     │                     │
│                      ├──────────────┤                     │
│                      │  Quotation    │                     │
│                      │   analysis    │                     │
│                      ├──────────────┤                     │
│                      │  Resolving    │                     │
│                      │   queries     │                     │
│                      ├──────────────┤                     │
│                      │ Post-tender   │                     │
│                      │  review and   │                     │
│                      │  feedback     │                     │
│                      └──────────────┘                     │
└─────────────────────────────────────────────────────────┘
```

5.1   Supply chain

This chapter looks at the acquisition of services and materials both from outside the contracting company and internally. The estimator or bid manager should play a key role in managing the relationship between the company and those people with whom the contractor needs to work. Contractors and sub-contractors have, traditionally, not had the best of relationships. However, the industry has moved forward, and in recent times, there have been efforts to remove the adversarial nature of the relationship and to replace it with a more conciliatory and harmonious one. It is now commonplace for contractors to 'single point' or even 'partner' with sub-contractors. The ethos of fair play, business integrity and team spirit is starting to prevail and the relationship between contractor and sub-contractor should remain harmonious from initial enquiry through to final account, continuing into the defects liability period.

Some of the resources required to complete a project may be available internally at a company and these need to be identified and costed. The dates during which they will be required should be established and checks made to ensure they are not committed elsewhere. The procedures identified for procuring services and materials externally should ideally also be used for those procured internally so that the departments providing these can keep adequate records and manage themselves effectively.

With the tendency for the major elements of the construction process to be carried out by specialist sub-contractors, there is less demand for material procurement. However, some contractors having a proportion of their own operatives and the numerous 'labour & plant only' sub-contractors, still make the acquisition of materials an important part of the supply chain process.

Consultants (architects, engineers, etc.) have traditionally been engaged and paid by the employer directly. However, with the increasing use of D&B it is now often the case that they will be engaged and paid by the contractor. The engagement of consultants is now often part of the contractor's procurement process, and must be managed effectively. Advice for the procurement of consultant services is covered in Chapter 11.

Many contractors also form a strategic alliance with their specialists. Such 'partnering' agreements typically give a contractor an undertaking by the specialist to provide an improved service and preferential prices in exchange for an undertaking by the contractor to give them a continuous flow of orders or enquiries. More sophisticated arrangements may incorporate an undertaking to make continuous improvements to performance and prices.

Partnering agreements with specialists and with clients commonly incorporate key performance indicators (KPIs) that give a quantitative measure of performance in key areas such as:

- quality of work or product in terms of finish and function, etc.
- achievement of programme objectives
- keeping to costs targets
- HSW&E standards
- considerate contractor policies
- training of staff and operatives.

Package identification

At this early stage, it is important to decide which packages the project can be divided into for sub-contracting purposes along with the extent of the design input that may be required for any of the packages. The extent of the overlap between packages needs to be assessed as this will have an impact on the ease with which (at a later stage) the contractor can make changes in the sequence or timing of the work and the co-ordination costs.

Supply chain management

Supply chain management encourages contractors to employ the same specialists on a number of projects, in effect reducing the numbers of specialists on their lists, in order to maximise the benefits of good team working. At the beginning of a working relationship, specialists will go through a learning curve as they become used to the procedures and staff of a particular contractor. Over time, relationships become closer and team working becomes more effective, not only between the contractor and a particular specialist, but also between specialists that regularly work together, to the point that a degree of integration will take place. Contractors and specialist can develop relationships in which great mutual benefit is derived from working together, which is reflected in a significantly improved service to clients. This is particularly effective on serial contracts or repetitive types of work such as building houses. Design and product compatibility is maximised, quality standards rise and costs fall.

5.2   Take off quantities

It is becoming common practice, even on large projects, for the client's team not to produce a bill of quantities for tender purposes. Where drawings and specifications are provided, it may be necessary for the contractor to take off their own quantities and perhaps prepare their own bills of quantities. Quantities enable materials to be purchased and labour-only sub-contractors to be employed. The use of quantities in pricing documents sent to sub-contractors ensures that their tenders are on an equal basis and eases evaluation of the works and cost comparison.

5.3   Preparation of enquiry documents

A list of components for each enquiry package is made by abstracting these from the tender documents. This list is then used as a tick sheet while assembling the package, as a record of what enquiry documentation was sent, and should be reproduced in the enquiry document (as a documentation log) so that the recipient can check that all the relevant components have been received. Any components which are to follow should be so described.

If bills of quantities are not provided with the tender documents (which is more usual these days) then the contractor must decide if quantities are to be prepared and the issue of any enquiries delayed whilst this is done, or if the enquiries are to be sent without bills of quantities, or indeed if a combination of each will apply.

There is no doubt that most recipients would prefer to receive enquiries with bills of quantities as these require far less time and resources to price than those requiring them to take off quantities and, therefore, they are far more likely to be priced and returned.

There is also a risk factor to consider as in the absence of a bill of quantity, the onus and control is with the sub-contractor/supplier to measure the work content correctly. If the sub-contractor/supplier makes a mistake in his measure then experience indicates that if the error is an overmeasure, the price will not be competitive. Conversely if the error is an undermeasure, then the price will be competitive, and the contractor may well use the figures submitted. In the latter case, once a sub-contract exists, then the subsequent loss is with the sub-contractor. However, in practice, the error generally comes to light after the contractor's tender has been accepted, and before the subcontract has been signed. Thus, the sub-contractor will invariably revise the price, leaving the contractor, by default, with the effects of the error.

It is understandable, given the foregoing, that most estimators would rather use bills of quantities with enquiries. If it is decided not to use bills of quantities with the enquiry, then there should be at least a list of headings against which a breakdown of a lump sum price can be given. This will allow the estimator to compare and evaluate various prices using the same list of headings and not have to base an assessment on the total price with no opportunity of interrogating it.

The use of standardised documentation and procedures assists at this stage in the methodical preparation of the estimate, and allows an interchange of personnel at any stage.

Lists and pre-selection

Contractors should maintain comprehensive records of suppliers and sub-contractors. These records should include:

■ details of past performance on site

■ previous performance in returning prices on time

■ extent of geographical operation

■ size and type of contract on which previously used

■ information concerning contacts

■ address, telephone and fax numbers

■ e-mail address and website (if they have these)

■ notes on quality assurance (QA) registration.

When operating in a new area, a contractor needs information concerning the local suppliers and sub-contractors. In this case performance should be verified from other external sources and any remaining information established from the supplier and sub-contractor concerned.

A questionnaire may be used to establish the resources and abilities of sub-contractors concerning:

■ area of operations

■ size and type of work

■ labour and supervision available

■ size and type of work previously carried out

■ references from trade, consultant and banking sources

■ insurances carried by the sub-contractor (if relevant)

■ confirmation of holding of relevant sub-contractor's tax exemption certificate (if relevant)

■ QA registration

■ health and safety record.

This questionnaire attempts to establish the supplier or sub-contractor's financial capability to undertake the work in question and to supply the materials and plant required. It is necessary to establish that all resources will be available to meet the requirements of the main contractor's programme.

Pre-selection will be necessary if dealing with unknown suppliers and sub-contractors. Bearing in mind the particular needs of the project, the contractor should ensure that the list of suppliers and sub-contractors who are invited to tender is comprehensive and that bids will be received. Pre-selection procedures should confirm that quotations will be submitted and establish that the particular supplier or sub-contractor has the necessary resources and desire to tender for the project. The procedures described in Section 5.2 should be applied if sufficient time is available. In any event, confirmation should be obtained from major suppliers and sub-contractors that they will be prepared to submit a bid, before enquiry documents are sent out.

It is also important to note that under CDM 2007 regulations, specific requirements have been placed on reasonable efforts to establish competency, particularly where design responsibilities are involved.

Works planning and method of construction

Sub-contractors and specialist suppliers often have particular skills and ways of working which make their role and contribution not only critical to the success of the project, but also difficult to plan at the tender stage. While, in theory, sub-contractors and suppliers are entitled to share the contractor's programming information, in practice it is common for contractors (misguidedly) to refuse to give the sub-contractor or suppliers anything more than a start and completion date within which to work. Where the sub-contractor or supplier has to interface with other sub-contractors or suppliers which likewise only have a start and completion date to work to, without any guidance on how the contractor sees the interface working, the resultant difficulties are notorious.

The contractor should thus involve all likely sub-contractors and suppliers in preparing the tender works programme and it is essential that this programme incorporates the sub-contractor's interfacing and periodic working patterns.

Tender works programming information should include:

- the anticipated start date for the main contract
- the anticipated start dates for sub-contractor, or delivery of materials
- the anticipated required completion date(s)
- the anticipated interfacing with others
- any anticipated periodic working
- key information of significance to the progress of the works
- details of any phasing and/or sectional completion requirements.

Suppliers should be given an indication of the rate of delivery, together with any requirement for the approval of samples prior to delivery to the site. Although sub-contractors will be provided with details from the contractor's method statement, which are relevant to their work, contractors may be reluctant to divulge information which could assist their competitors. However, in relation to any sub-contractor or supplier, if the contractor is unaware of, or, unable to specify any

- required interfacing
- periodic working or
- key information of significance to the progress of the work

that information should be requested from the sub-contractors and/or suppliers and incorporated into the tender works programme.

Other issues

Any other issues to be conveyed to the prospective suppliers and sub-contractors should be included. These may include environmental issues such as sustainability, Building Research Establishment Environmental Assessment Method (BREEAM) requirements, non-discrimination policies, health and safety procedures and induction details, attendances provided, etc.

Figure 5.1 Typical enquiry letter to sub-contractors.

[Date]

To [sub-contractor]

Dear Sirs,

[PROJECT TITLE]
[SPECIALIST TRADE PACKAGE]

We invite you to tender for the above work and enclose the following documents which will form the basis of your offer:

Site plan drawing number:	A/100
Preliminaries pages:	1/1-12
Specification pages:	2/23-32
Bill pages:	4/3-5, 7, 11-13
Drawings numbered:	A/201-205, 304, 305
Daywork schedule:	7/1-3
Health and safety plan:	
Form of tender:	

Your form of tender, priced bills of quantities and daywork rates must be delivered to this address to arrive no later than 12 noon on [date].

The form of sub-contract will be DOM/1 incorporating all relevant published amendments, and the following:

Payments:	monthly
Discount to main contractor:	2.5%
Fluctuations:	firm price
Liquidated damages:	£950.00 per week
Retention:	5%
Method of measurement:	SMM7
Defects liability period:	six months

We will provide all sub-contractors with water, lighting and electricity services near the work and common welfare facilities on site. Sub-contractors will be required to provide the following services and facilities:

1. Unloading, storing and taking materials to working areas.
2. Power and fuel charges to temporary site accommodation.
3. Clearing-up, removing and depositing in designated collection points on-site all rubbish and surplus packing materials.
4. Temporary accommodation and telephones.
5. Day-to-day setting out from main contractor's base lines.

If you have any queries or wish to arrange a site visit please contact the estimator for the project [name].

Would you please acknowledge receipt of this enquiry and confirm that you will submit a tender in accordance with these instructions.

Yours faithfully,

Enquiry letter

It is important to convey the nature of the project, together with any restrictions (method, timing restrictions and requirements, access, working hours, noise, etc.) to the recipient. The enquiry letter should clearly convey what is required, the procedure for raising queries, the situation regarding alternatives, main contract conditions that will be mirrored in any subsequent order, and when and in what form the price is to be returned. A typical enquiry letter is shown in Fig. 5.1. (This refers to, and should be used in conjunction with, the documentation log). Note that, where relevant, standard information, such as health and safety induction procedures, provision of normal attendances, and QA procedures can be dealt with by reference to the contractor's standard documents which can be included in the documentation log. Recipients who trade on a regular basis with the contractor can be assumed to hold copies of the latter and this can be stated, with the option to request copies, if this is not the case. Obviously new recipients should receive them with the enquiry.

Offer, quotation or simply prices

The five essential characteristics of a contract are:

1. offer

2. acceptance

3. certainty of terms

4. intention to create legal relations

5. capacity to contract.

It is not the remit of this Code of Practice to deal with contract law; however, it is worth beginning this section with a short explanation of the legal interpretation of each.

Items 4 and 5 rarely create difficulties in construction contracts. As in commercial relationships an intention to create legal relationships is inferred and there are very few companies that are devised without the power to contract for building works. However, be aware that, for example, a charity set up to provide an educational benefit may not be empowered to develop its land speculatively for sale. Certainty of terms is also rarely in question unless either the nature of the work contracted for or the price remains undefined, or a bespoke form of contract or one, amended by special terms, is used.

As to offer, a distinction must be made between an offer to perform ('offer') and an invitation to receive offers to perform ('invitation to treat'). The former can be matured into a contract by acceptance but the latter cannot. For example, if an employer advertised for tenders to be submitted, that would constitute an invitation to treat and no contract would be formed simply by submitting a tender. In that example, the tender is the offer, which can only be accepted by a specific acceptance of it.

A schedule of prices submitted by way of an offer can be accepted in the same way a quotation of a fixed price can be accepted and both can mature into a contract. An 'offer', once accepted, produces a binding obligation to carry out work or supply materials at the prices and under the terms stated in the 'offer'. It is the first step in the contractual process to create a binding contract. 'Offer' plus 'acceptance' = 'contract'.

A 'quotation' is merely a statement of price and **not** an offer. It is referred to in legal terms as an 'invitation to treat'. The offer is actually made up by the contractor, who effectively 'offers' the order to the sub-contractor/supplier. The sub-contractor/supplier then 'accepts' by either signing a sub-contract, commencing work or delivering materials.

Generally, an acceptance must relate to the work for which an offer was submitted. A conditional acceptance, or a letter of intent referable to only a part of what was offered, is not an acceptance; it is a counter-offer which can only mature into a contract by acceptance.

Care should be taken to read any small print on an offer to carry out work, as the conditions imposed may not be acceptable to the client, and may not be able to be passed on under the main contract.

Selection of recipients

Depending on the individual enquiry and the quantity of resources necessary to provide a competitive price, the contractor needs to decide how many enquiries are issued and to whom.

It should be a fundamental rule that enquiries are only issued to recipients who are likely to acquire the business if the contractor's tender is successful, and who are likely to submit competitive prices. A minority of sub-contractors and suppliers are reluctant to provide competitive prices at the main contract tender stage, and rely on being able to make a further approach to the successful contractor once the main contract has been secured. Furthermore, a minority of contractors will be inclined to send out further enquiries once the main contract is secured, because they believe that they will obtain more competitive prices now that they are in a position to place an order. Both these practices can be deemed to be unfair to the very sub-contractors and suppliers whose prices enabled the contractor's tender to be successful in the first place.

The number of enquiries, and the list of recipients, should take into account the cost and resources required by recipients to provide competitive prices, and weigh this against their likelihood of acquiring the subsequent business.

If the enquiry is for a major element, requiring the recipient to invest considerable costs and resources in the preparation of a competitive price, the contractor may consider a 'single-point' solution. This means that the contractor only sends one enquiry to a favoured recipient on the understanding that if the contractor's tender is successful then that recipient will be awarded the business. In return for this favoured status, it is understood that the recipient will endeavour to produce the most competitive price, and will also examine potential alternatives that may provide the contractor with a competitive edge. This approach is common in design and build tendering, particularly in respect of mechanical and electrical services, and allows the prospective sub-contractor to be an integral part of the contractor's design team throughout the tender period.

Types of sub-contractors

Essentially, there are two types of sub-contract: those selected by the contractor (domestic sub-contractors) and those selected by the employer with whom the contractor is instructed to sub-contract (nominated sub-contractors). Listed sub-contractors fall into the first group as, although the employer provides a list of companies from which a selection is made, it is the contractor who makes the selection.

Nominated sub-contractors are normally the subject of PC sums, which means that the employer or his consultants will issue enquiries for them and make the selection. The contractor should identify in its tender works programme, nominal dates for receipt of instructions for the appointment of nominated sub-contractors including, where appropriate, nominal periods for design, fabrication, delivery and installation.

In regard to the content of sub-contract works, there are four categories:

■ Those who provide design (and sometimes off-site fabrication), all necessary labour, plant and materials.

■ Those who provide all necessary labour, plant and materials for work only.

■ Those who provide the labour and plant resources, with the contractor providing materials for work only.

■ Those who simply provide the labour resource, with the contractor providing both plant and materials for work only.

5.4   Quotation receipt

Sub-contractor's enquiries

Some sub-contractors will return prices as soon as they get the enquiry, some during the return period, others just in time to meet the return date deadline, and a few will not respond at all. A few will respond declining to price, and will usually give a reason (e.g. too busy, estimator on holiday, etc.).

It is important to contact those who have not responded in order to discover why. If there has been a simple oversight, then prices may still be obtained in time to be considered. Perhaps the database needs to be adjusted if the recipient has relocated and the enquiry may have gone to an earlier address. There may be a grievance, from a previous dealing, which has influenced the recipient not to price. If so, then this may need to be addressed.

The database should keep a record of enquiry responses, and those who have declined to submit prices should be checked against their previous response record. Again, if there is a history of declination, then they should be contacted and the matter discussed. Sending enquiries to sub-contractors who traditionally do not return prices is a futile waste of resources, and results in poor price coverage. If there is a short period between the enquiry return date and the tender settlement, then a telephone call to key sub-contractors prior to the enquiry return date to check on progress would be in order. This kind of contact can produce interesting feedback on the specific elements of the project from the sub-contractor's viewpoint, and also important information regarding the activity of other contractors who may have sent enquiries to the same sub-contractor.

Queries raised by sub-contractors should be dealt with as a matter of urgency, as any delay will be cited as the reason for not returning prices by the required date.

Supplier enquiries

All of the previous section regarding sub-contractors can be considered as relevant here, however, it should be remembered that those providing materials as opposed to sub-contract services are often likely to be less interested in the nature of the project and its restrictions. Often, also, the person dealing with the price submission may come from a clerical background rather than from a specialist trade or profession.

Quotation receipt

On receipt, the date should be entered on the enquiry record, and basic information such as address, telephone number, fax number and e-mail address should be

checked against those held on the sub-contractor/supplier database and amended as necessary.

The onus to check and verify

The onus is on the contractor to read, check and verify what is received from sub-contractors and suppliers. If a price submitted is erroneous, the sub-contractor/ supplier can, and invariably will, refuse to complete the contractual process, leaving the contractor with the consequences. Even if a *bona fide* offer is provided, there will invariably be a period of time before the contractor is able to accept it, leaving the way clear for it to be withdrawn, or a revised offer to appear.

5.5   Quotation analysis

It is vital that submissions are read thoroughly and understood before they are used to underpin the estimate. It is not unknown for specialists, in an attempt to make the price seem more competitive, to hide vital components in other parts of the submission where they appear as optional extras to the stated price, whereas in reality they are necessary and not optional.

With particular regard to material prices, it is important that the suppliers with the most competitive prices are contacted, to confirm that they noted all the relevant restrictions on delivery to site. It is also important to check that the supplier can meet important delivery dates and supply the materials in the quantities needed. All of this will be stated on the enquiry, but it is not always read by the recipient.

Obviously, breakdowns of price should be checked for correctness, and this is often done while entering the rates into a computerised estimating system, by comparing the relevant page totals.

Analysis and comparisons

If the basis of the original enquiry was drawings and specifications, with no bills of quantities, then the price will be a lump sum. In this case, the list of headings (referred to in Section 5.2) will at least allow comparison between prices received against the list headings sent with the enquiry. If several prices have been received, it will quickly be apparent if one is lower or higher than normal in any specific area, and queries can then be raised.

Bills of quantities allow far more detailed analysis and comparison. Even if only one price is received, then with the benefit of experience and/or reference to prices received on previous projects. the estimator will quickly gauge its adequacy.

If several prices are acquired with bills of quantities breakdowns, then the individual rates can be compared, and those higher or lower than normal, highlighted and subsequently queried.

All good computerised estimating systems will have a facility to do these comparisons. They will also have facilities to display a list of comparison summaries in report format for use at the settlement stage.

Material prices are dealt with in much the same way as sub-contract prices. However, they tend to be entered into comparison forms, which allow a degree of mixing. Whereas it is normal to have one sub-contractor to carry out an entire section of work, it is also normal to procure materials from several suppliers, e.g. doors, hinges, ironmongery, door linings and architraves could all come from separate suppliers. Understanding which materials can be safely mixed, and which suppliers will be prepared to provide a selection from a larger submission without requesting a review of prices, is an area where experienced judgement is required.

5.6 Resolving queries

This is an area where caution needs to be tempered with integrity. Here we are looking at any queries the contractor may wish to raise regarding the quotations received. There is a difference between querying specific rates or matters referred to in a particular submission and divulging privileged and confidential information provided in other submissions. Whereas it is in order to advise that a particular rate seems low or high and that confirmation of its correctness, or some form of amendment, is required, it is not appropriate to give details of rates and prices provided by others which have led to the query being raised.

A record should be kept of queries raised, with names of individuals spoken to, together with dates and times. The prices submitted should be flagged as having an outstanding query until the query is resolved, and all good computerised estimating systems have this facility. A list of outstanding queries should be kept, with details of the specific query appended to, or written on, the individual submission. Once resolved, queries are then crossed off the list, the revised prices or amended submission appended to the original submission, the rates amended accordingly, and the outstanding query flag removed.

Gap analysis

There will be instances where one submission has elements that have not been priced and in order to compare it with others, the gap needs to be initially identified and filled by inserting a value. The way in which this value is decided will depend on the case in question, and there are no rigid rules that can be applied.

If the gap is simply a rate that in error has not been priced, then an average of other rates received can be used for comparison purposes. However, if the subsequent price is of interest and may be used in the estimate, then the sub-contractor/supplier should be contacted and a rate obtained.

If the gap is due to one sub-contractor not being prepared to carry out a particular section of work, then the estimator will need to decide the cost (including the risk) of using another sub-contractor, or directly employed operatives, to do this and use that cost to fill the gap.

5.7 Post-tender review and feedback

Updating the sub-contractor/supplier database

The sub-contractor/supplier database requires constant maintenance and it is good practice to check incoming enquiry returns for areas which need amending.

New sub-contractors and suppliers may have been uncovered during the tender process, and these should be fully checked and verified, and full details entered onto the sub-contractor/supplier database.

A database is only reliable if it is kept up to date.

Communicating the result

Once the outcome of the contractor's tender is known, it is good practice for the inviting party to send a result circular to all those who submitted prices, thanking them for their submission and advising them of the outcome.

If the contractor's tender was successful, the result circular should contain contact details for the person who is to place orders.

It must be borne in mind at all times, that those who submit prices do so with the intention of obtaining future work, and the circulation of this information as soon as it is known, particularly where the contractor's tender is unsuccessful, is vital to aid them in that quest.

Sub-contractors' forum

Many contractors hold regular sub-contractors' forum meetings, where a selection of sub-contractors are invited to attend and voice their views on the contractor's working and procurement methods. These are normally attended by the contractor's senior management, and both positive and negative feedback is welcomed and recorded. This feedback information is invaluable in formulating future policies in relation to working and procurement methods.

6 Pricing the works

	Pricing the works — Chapter 6	
Inputs	Calculation of net unit rates	Outputs
	Estimating all-in rates for labour	
	All-in rates for plant	
	Plant quotations	
	Materials quotations	
	Sub-contract quotations	
	Unit rate calculation	
	PC and prov. sums	
	Preliminaries and project overheads	

6.1 Introduction

The contents of tender submission documents vary considerably depending on the form of contract and documents used to define the works. Here the pricing of bills of quantities for building work is considered to demonstrate the analytical or 'bottom up' estimating process. This includes the building up of unit rates, pricing of project overheads and dealing with PC and provisional sums.

The bills of quantities are normally divided into five sections:

1. preliminaries

2. preambles (specification) – not priced

3. measured work

4. PC and provisional sums

5. summary.

In order to avoid confusion in calculating and analysing an estimate, it is recommended that PC and provisional sums should form a separate section at the end of the measured work part of a pricing document. Since this is not always the case, estimators must carefully check all the bill pages, including preliminaries, to ensure that the written-in sums are incorporated into the final tender amount. In this chapter we have looked at pricing the measured work first, followed by PC and provisional sums, before looking at the preliminaries. The calculation of a contractor's head office cost allocation and level of profit to add, is covered under tender settlement.

Computer input

Much of the estimator's analysis of costs is made using a computer model of a bill, which is entered in a number of ways, as follows:

- taking off quantities directly from drawings

- using a scanner to transfer written bills of quantities to a text file

- manual entry of item references, descriptions (if required), quantities and units

- bills of quantities are available on disk or by e-mail for immediate incorporation into a computer-aided estimating system.

If a contractor has the suitable software and can access the bills of quantities with full descriptions, enquiries to sub-contractors can be extracted and printed by the computer once the items have been coded by the estimator. For example, all the piling items may have a sort code S01 and metalwork S02, which are also the references used for sub-contract enquiries on the sub-contractor's enquiry form (Fig. 6.1). The computer can then find all items in a trade, print the relevant pages from the bill and take names and addresses from a vendor database.

The checklist may be used to ensure that the computer bills accurately represent the written document. The most important check is that all the applicable items are priced and quantities are correct. If an estimator prices the work items directly from the computer (without reference to the written document), then descriptions will be checked, particularly dimensions forming part of the text.

There are many potential hazards that an estimator must look for before passing documents to computer assistants. The bills of quantities may need some simple modifications in order to comply with the estimating software being used. Problems often occur in the following ways:

- page numbers can be difficult for a computer system to accept, such as 12A/A/D1: in this case the estimator could re-number the bill pages with a simpler reference such as 12 .1, meaning bill 12 page 1

- some bills have work items and collections on the same page; again many computer systems need different page numbers for collections and summaries

- any missing item references must be inserted by the estimator.

Figure 6.1 A specimen sub-contractor's enquiry form: quantities abstract and analysis.

	QUANTITIES ABSTRACT AND ANALYSIS		Tender number

ref.	description	quantity	unit	output hrs/unit	total time		gang size	duration wks
					hrs	man wks		
3/3	Formwork to foundations	1126	m²	1.75	1971	43.79	4	11
3/5	Bar reinforcement to founds	13.5	t	25	338	7.50	2	4
3/1	Concrete in foundations	200	m³	1.65	330	7.33	4	2
3/8	Blockwork below dpc	255	m²	0.75	191	4.25	2	2
3/14	Structural steelwork	55	t	15	825	18.33	4	5

6.2   Calculation of net unit rates

Principles

In calculating unit rates for inclusion in the bills of quantities, careful consideration must be given to every factor which may influence the cost of the work. Where work has been measured in accordance with a standard method of measurement (for example SMM7), then the definitions and coverage rules should be clearly understood as these will affect the pricing.

For example, the measurement of a drainage trench is deemed to include any necessary earthwork support, compaction, backfilling with excavated material and disposal of surplus soil. This would not be obvious just from reading the description of the item of work for the trench, but a contractor should make allowances for them in their unit rate.

Unit rates for measured items in the bills of quantities (excluding preliminaries) consist of any or all of the basic elements:

- labour

- plant

- materials

- sub-contractors

- overheads (site and head office)

- profit.

The addition for overheads and profit is often part of the tender settlement process and for clarity, is covered here.

It is recommended that each element of the unit rate is analysed and estimated separately.

Pricing strategy

The estimator needs a clear strategy for pricing the bills of quantities. It is unlikely that items will be priced in the order they appear in the bills because a better understanding of activities can be gained by pricing one trade at a time. This is becoming increasingly popular with computer estimating systems which readily sort the bill items into trade order (similar items). Computers also allow resources to be entered either through a resource build-up screen for each item or with the aid of a spreadsheet-type comparison system, where similar trade items can be viewed in a single table.

If it is known that quotations for materials will be delayed, the estimator can price labour and plant first, and return to part-priced items later when quotations are available. On the other hand, 'typical' materials prices may be used during the pricing stage. Computers allow late adjustments to be made and all affected items will be changed.

Composition of net unit rates

A net unit rate for an item of work is built up in three distinct stages listed below.

Stage 1   The establishment of 'all-in rates' for the key items that will be incorporated. These include:

- A rate per hour for the employment of *labour.* Different rates will be established for the different categories of labour that will be used on a project.

- An operating rate per hour (or per day, per week, etc.) for an item of *plant.* This is for plant supplied with, or without, operator and the rates are established from the contractor's own data or from quotations received from plant hire organisations.

- A cost per unit of *material* delivered and unloaded at the site. This involves comparison of the various quotations received for materials and the selection of one of these for use in the estimate.

The costs of labour and some plant items are established after visits to the site and consultants. It is also necessary to complete the tender works programme and method statement before finalising these prices.

Stage 2   The selection of methods and production standards from the contractor's database or other sources. These standards are then used in conjunction with the all-in rates calculated in Stage 1, to calculate net unit rates which are set against the items in the bills of quantities. Alternatively, rates received from sub-contractors are used.

Stage 3   The incorporation of rates from specialist trade contractors, including those offering labour-only services, either producing the whole or part of a rate.

The calculation and addition of project overheads is a separate and subsequent operation dealt with in Section 6.10, and the preparation of reports for consideration by management is considered in Chapter 7.

6.3   Estimating all-in rates for labour

Items for consideration in all-in rates for labour

Labour costs arise in two areas: costs associated with the Working Rule Agreement and certain overhead costs incurred by the employer. Other costs, which will be variable and are specific to a project, or some time-related costs, will be contained in the project overheads.

Labour costs normally contained in the all-in rate			
	a.	Guaranteed minimum wages	*basic rates in NWR 2 for each class of operative*
	b.	Contractor's bonus allowance	*bonus used to retain operatives which is not self-financing*
	c.	Inclement weather allowance	*normally included by paying full weekly wages*
	d.	Non-productive overtime costs	*NWR 7 gives the rules for calculating overtime rates*
	e.	Sick pay allowance	*NWR 16 gives amount and qualification rules*
	f.	Trade supervision	*proportion of non-productive time by supervisors often considered in project overheads*
	g.	Working Rule Agreement allowances	*NWR 3 and 4 list extra payments for labourers engaged in particular activities*

	h.	CITB training contributions	*normally levied at 0.25% of entire payroll, and 2% for self-employed and labour-only sub-contractors*
	i.	National Insurance contributions	*employer contributions are percentage of weekly earnings depending on which of six bands apply*
	j.	Holiday credits	*annual and public holidays*
	k.	Tool allowances	*given in NWR 18*
	l.	Severance payments	*statutory scheme*
	m.	Employer's liability insurance	*often considered in project overheads*
Labour costs normally contained in project overheads	n.	Daily travel allowance	*NWR 14 provides a scale of allowances for daily travel one way. Some employers provide transport, which reduces the allowances*
	o.	Periodic leave and lodging allowance	*travelling and lodging allowances are given in NWR 15*
	p.	Supervision	*usually priced using tender works programme*
	q.	Attraction money	*needed in remote areas or sites which might experience shortages of local labour*

General

It may not be possible to determine all of these factors with accuracy at an early stage of estimating, particularly as the volume of labour is not quantified. There are benefits in separating them from the calculation of all-in rates and for necessary allowances to be made in the later stages of the estimating process, in the project overheads.

It is a matter of opinion, and company preference, where many of these items are priced. The distinction drawn between items set out in all-in rates and project overheads, should not be regarded as mandatory. The important consideration is that due recognition must be given to all items to establish the true costs involved and that adequate allowances are made in the estimate.

All-in rates are built up on a weekly or annual basis, or the time period relating to a particular contract. In this Code, the all-in rate is calculated on an annual basis.

Calculation of an all-in hourly rate for labour

There are three stages in the calculation:

1. Determine the number of working hours which an operative is expected to work during the one-year period.

2. Calculate the cost per year for wages and the cost of each item used from the list above.

3. Summarise the individual costs obtained in (2) and calculate the all-in rate per hour by dividing the total costs in (2) by the number of hours in (1).

Alternatively, for a very large project, it may be desirable to make a special calculation based on the anticipated construction period.

Examples of this are included in the Appendix at the end of this chapter.

1. It is emphasised that the calculations in the examples are for guidance only.

2. Details vary according to:

   ■ the actual trade of the operative (whether craft operative, labourer or mechanical plant operator)

   ■ the firm

   ■ the area

   ■ the industrial and legal conditions in force at any time.

3. In the example, the calculation is based on a one-year employment period and could apply to all projects for which tenders are to be submitted. It is for a craft operative.

4. Amendments must be made each time there is a variation in the cost of one of the factors included in the calculations, or when further factors are introduced. Alterations must be made as soon as variations are promulgated, although there may be a period of time before they come into effect.

5. The whole calculation should be revised regularly.

Determination of hours worked

The number of hours worked during the calendar year, i.e. January to December, will depend on the hours worked per week during the summer and winter periods, with adjustments for annual and public holidays. The hours worked will vary between different companies and some variation in hours can also be expected between firms operating in the north and south of the UK, due to the available amount of natural daylight hours in the winter period. Local customs and availability of labour also affect the number of hours worked. A company may agree to work hours suitable for a particular type of contract, due to special requirements of the employer or by special agreement with its employees.

The Working Rule Agreement states that a 39-hour week should be worked throughout the year as follows:

■ Mondays to Thursday: 8 hours per day

■ Fridays: 7 hours.

For calculation purposes only, the hours used are typical of those worked on many sites, although in the 'summer period' (defined in the Working Rule Agreement as 1 April to 30 September) it is common to assume that a longer working day can be achieved throughout British Summer Time, assumed to be 30 weeks long.

An operative's annual and public holidays have been taken as listed in the Appendix and the total number of days are as agreed in the Working Rule Agreement. In some regions the actual day to which the holiday is allocated will vary according to local tradition.

Production standards and other considerations

The records of cost and outputs achieved on similar work from previous projects is a major source of information used in estimating. This information arises from records of resources used on site or from work study exercises to establish standard production rates. It is important to remember that the cost or output depends on many variables and attention should always be paid to the conditions that prevailed at the time when the particular recorded cost or output was noted, and consideration given to the levels of incentives which were used to achieve the particular standard. These conditions must be compared carefully with those expected to be encountered on the project under consideration. Differences between the estimated and actual cost or output on previous projects should be analysed and any obvious conclusions noted. Adjustments must then be made to update the estimating data.

When a particular type of work is being considered for the first time, there will be no previous cost or output records for guidance. Trade specialists should be consulted whenever possible and technical information from outside sources may have to be used. Information can be provided by manufacturers either through printed literature or technical representatives. Caution will be needed when using data from external sources.

Proper allowances must be made for any learning curve associated with new types of work and for incentive payments, either by increasing the all-in labour rate or by using an appropriately modified production standard. It may be necessary to adjust for contractor's bonus which has been included in the all-in labour rate. This can be adjusted later in the project overheads, if necessary.

Labour element

Labour costs are estimated on the basis of the all-in hourly rates previously established. It is recommended that 'gang costs' should be used for some trades in preference to individual hourly rates. Typically, an effective rate is built up for a member of a concreting gang, or a bricklayer rate that will include a proportion of labourer's time. However, it is also reasonable to price all trades on a normal hourly rate basis provided that an allowance for attendant ancillary labour is added to the established all-in hourly rates or added as an item of general labour in the project overheads stage.

It is usual to express outputs as 'decimal constants' such as 1.50 hours to lay a square metre of brickwork. This is because computer systems conventionally expect estimators to rate items by inputting a resource code and a quantity (see the introductory figure to this chapter).

Contractors and trade specialists assemble tables of data for use by their estimators as a guide to basic outputs. Many factors affect the time allowed for an operation or item and careful consideration must be given to each of these factors, enabling the time allowed to be as accurate as possible.

The drawings, specification and bills of quantities should be carefully examined to determine:

- the quantity of work to be done

- what allowance is needed for compaction, overbreak, batters, etc.

- quality of finish and standard of workmanship required

- whether operations are repetitive

- whether excessive or detailed setting out will be required

- degree of accuracy and tolerances required

- whether the design of the work is intricate or straightforward

- whether any special skills will be needed

- whether any special construction sequence is necessary

- whether the operation is likely to be within the experience of existing staff and operatives, or whether special instruction or training will be needed, or whether there will be a need to engage specially trained personnel

- position on site in which the work occurs

- accessibility of work

- height or depth of work

- any double handling of materials

- the weight of specific items

- restrictions in working, such as secure areas, safety

- shift times

- environment, such as hot/cold/exposed areas of work.

The tender stage construction programme and method statement will indicate:

- time available for activities on the site

- time of the year when work is to be carried out and the likely seasonal conditions to be encountered

- whether work will be continuous or intermittent

- any restrictions which might affect normal working

- degree of interdependence of trades and operations

- facilities available for use by domestic and nominated sub-contractors as items of general attendance

- pattern of production and the likelihood of achieving maximum possible rates

- resources needed, such as the relative proportions of supervisory, skilled and unskilled operatives required, and recommended gang sizes

- extent of mechanisation envisaged and method of unloading, storing, handling and transporting materials.

The visit to the site and locality will have shown:

- physical conditions and any restrictions likely to be encountered

- site layout, operating storage and unloading facilities

- likely skill, experience and availability of local labour.

The production standards to be used by the estimator in establishing labour costs will take into account the company's existing data and experience, as well as any circumstances associated with the project information, tender works programme, method statement or visits to the site and consultants, which indicate that these standards should be modified in some way. The estimator must record any special factors which lead to alteration of production standards to any considerable extent, in the Estimator's Summary and Report for consideration at the settlement stage.

Figure 6.2 A specimen scaffold schedule.

	SCAFFOLD SCHEDULE	Tender number	

Ref.	Type	Description	Quant.	Programme	
				From	To
1	Ext independent scaffold	South wing three elevations		wk 6	wk 26
2		Part north wing around stairtower		wk 6	wk 26
		Note: boarding for access to windows and roof parapet			
		Note: allow for fixed ladder access to both areas			
3	Internal scaffolding	To lift shaft		wk 12	wk 18
4	Temporary roof	South wing only		wk 8	wk 24
		Note: waterproof sheeting down to parapet level			
		Note: allow for temporary rainwater installation			
5	Loading platform	To south wing at parapet level 4.00 × 5.00 m		wk 6	wk 26
6	Hoist towers	Not required			
7	Roof edge protection	Electrical sub-station only		wk 24	wk 29
8	Debris netting	North wing stairtower scaffolding		wk 6	wk 26

Related documents	Drawings
	Specification
	Bill pages

Pricing the works

6.4   **All-in rates for plant**

The contractor's plant requirements will be established in the method statement and programme. They will establish the basic performance requirements of the plant and in many cases will have identified specific plant items needed for the works. The duration for which the plant is needed on site will be established from the tender works programme. The estimator must first compile a 'schedule of plant requirements', listing the type, performance requirements and durations. This should be separated into:

- mechanical plant with operator

- mechanical plant without operator

- non-mechanical plant.

A note must be made on the schedule of additional requirements associated with a particular item of plant, which must be provided by the contractor. A power supply for a tower crane, for example, could be a significant additional cost and temporary access roads for erection purposes may be needed, together with foundations.

Further details are necessary for certain non-mechanical plant. A scaffolding schedule must be drawn up by the estimating team in order to provide scaffolding contractors with a clear list of requirements. There are seldom work items in a bill of quantities for temporary works, although the preliminaries may give specific requirements such as temporary roof structures or bridging scaffolding to span low level obstructions. A typical scaffolding schedule is shown in Fig. 6.2.

The analysis of quotations received for plant will be set out in the plant comparison form (Fig. 6.3) and any additional factors to be priced identified. Allowances must be made for additional matters associated with the plant. In considering the total costs of plant, decisions must be made concerning the following items:

- the manner in which time-related charges and fixed charges will be accommodated, i.e. delivery, erection and removal charges could be spread across the duration of hire and added to a weekly rate or, alternatively, shown separately as a fixed charge, separate from time-related costs in the project overheads

- rate of production likely to be achieved by the plant, bearing in mind the specific requirements of the project, the season of the year, and in the case of excavation work, the ground and water conditions

- the continuity which can be expected for any item of plant and the likelihood of achieving a high production rate; it is unlikely that outputs quoted by manufacturers can be attained

- average output, making due allowance for intermittent working, site conditions, seasonal effects and maintenance.

In all cases, the tender works programme requirements must reflect these conditions.

In establishing the costs of plant, allowances must be made for additional matters associated with the plant. These include:

- divergence or discrepancy from the contractor's enquiry in the quotation which is being considered

- delivery, erection and removal charges if applicable

- fuel costs, if applicable

- availability of power for electrically operated plant; consider the need for a temporary sub-station or generators

- the effect and cost of maintenance and consequent down time of plant

- special provisions needed for unloading and loading plant

- temporary access roads, hard-standings or temporary works required for the plant

- weight restrictions which may affect the plant or its use

- whether any special insurances are needed for the plant, such as responsibility for the plant during delivery and erection

- consents required for the use of plant on or over adjacent land

- contractor's attendant labour requirements; banksmen are particularly important

- safety measures that are required

- supporting equipment needed to operate plant, such as crane slings, chains, skips, cages, etc., associated with lifting equipment, hoses, breaker's points, etc., associated with a compressor – these items may be separately priced in the project overheads

- allowance for damage, repairs and replacement parts chargeable to the contractor

- minimum hire charges.

The estimator must decide which items will be accommodated in the all-in rate for plant, where plant is to be allocated against unit rates, and which items are to be allocated in the project overheads.

Allocation of costs

When mechanical plant is used only on specific and limited operations (such as excavation and soil disposal), there is little difficulty in allocating the costs of the plant to specific items measured in the bills of quantities, taking into account the various factors noted above.

However, when an item of mechanical plant serves a number of trades or operations (e.g. a crane or hoist or a concrete mixer, which is used for concrete work and also brickwork and drainage work), then the allocation of its cost to measured items can only be made on an arbitrary basis. When the cost of an item of plant is associated with time on site rather than with specific items of measured work (e.g. pumping operations), then such items cannot reasonably be allocated against measured work.

In such circumstances, the cost of such plant, together with time in excess of productive output, must be included in the project overheads, rather than spread in an arbitrary manner over measured rates.

There are many examples of resources which are difficult to allocate with one cost category, such as falsework to support soffit. Formwork may be included in the material or plant element of a unit rate or equally can be assessed separately in project overheads as temporary works. For building work, the cost of falsework is commonly in the net unit rate; but this is not the case for civil engineering where all supporting equipment and structures tend to be linked to a resourced short-term programme and temporary works calculations.

6.5   Plant quotations

The contractor must consider the intended method of working and any tender works programme requirements, in the specification for plant. Turnaround of equipment and striking time will dictate the amount of formwork and support and access equipment needed. A balance must be drawn between speed of operation and economy in establishing plant needs, and all must be clearly reflected in the plant enquiry.

A list of plant suppliers must be established from companies who can meet the project's requirements. The options available for obtaining plant include:

■   purchasing plant for the contract (in accordance with company policy)

■   hiring existing company-owned plant

■   hiring plant from external sources.

Purchasing plant for the contract

The decision to purchase plant for a particular contract is taken by senior management. Such a decision requires knowledge of plant engineering and will be made in accordance with the accounting policy of the company. Purchasing of plant is outside the scope of this Code, but for guidance purposes only the following general factors must be considered when plant is to be purchased for a project and sold on completion:

■   purchase price less expected resale value after allowing for disposal costs

■   return required on capital invested

■   cost of finance

■   cost of maintaining the plant and associated overheads

■   stock levels of spares

■   the company's policy on depreciation

■   likely working life of the plant

■   cost of insurances and taxes, e.g. road fund tax

■   any tax or depreciation allowances that are available

■   availability and cost of plant outside the company

■   accessibility of the site in relation to company depot and servicing centres.

The manner in which the cost of such plant is charged subsequently to a contract will depend on the accounting policy of the company.

Hiring existing company-owned plant

When plant is already owned by the company, the estimating department will be provided with hire rates at which plant will be charged to the site. The following list should be regarded, as guidance only, to the items which must be considered in building up hire rates for company owned plant:

- a capital sum based on the purchase price and expected economic life (this will vary according to the company's accounting policy)

- an assessment of the costs of finance

- the return required on capital invested

- grants and financial assistance available when purchasing the plant

- administration and depot costs

- costs of insurances and road fund licences

- maintenance time and costs and also cost of stocks needed for maintenance purposes.

Hiring from external sources

Where company owned plant is not available, enquiries must be sent to external suppliers for the plant required.

Enquiries for items of plant must either be sent specifying particular machines and equipment that are needed or specifying the performance required from the item of plant. For example, 'tracked excavator with backacter, required to excavate trenches to a maximum depth of 3.00 m, width of excavation 1.00 m, 360° slewing required'.

Enquiries must state:

- title and location of the work, and address of the site

- specification of the plant or work to be done

- anticipated periods of hire with start date on site and duration required

- means of access, highlighting any restraints or limitations

- any traffic restrictions affecting delivery times

- anticipated working hours of the site

- date the quotation is required

- period the quotation is to remain open

- whether fluctuating or firm price required, the basis for recovery of fluctuations and the base date when formulae are used for the recovery of fluctuations

- discounts offered

- the person in the contractor's organisation to be contacted regarding queries.

In addition to the basic hire charge per hour or week, the enquiry must seek to establish:

- cost of delivering and subsequent removal of plant from the site on completion of hire

- cost of any operator, over and above the basic hire charge, if provided by the hiring company. However, if provided by the contractor, the estimator must produce a built-up rate for the operator's costs

- whether the hire rates quoted include for servicing costs; if not, the costs and timing of servicing must be established

- any minimum hire periods applicable to the plant and the extent of any guaranteed time

- cost of standing time and insurance costs if the plant is retained on site, and not working for any reason.

6.6 Materials quotations

Enquiries to suppliers of materials should state:

- title and location of the work, and site address

- specification, class and quality of the material

- quantity of the material

- likely delivery programme, i.e. period during which supplies would be needed with daily or weekly requirements where known; where small quantities are to be called off from a bulk order this should be clearly stated

- means of access, highlighting any limitations or delivery restraints, and any traffic restrictions affecting delivery times

- special delivery requirements such as palletting or self-unloading transport

- date by which the quotation is required

- period for which the quotation is to remain open

- whether fluctuating or firm price required, the basis for recovery of increased costs and the base date when a formula is used for calculation of fluctuations

- discounts required

- the person in the contractor's organisation responsible for queries.

The contractor has a responsibility to ensure that suppliers:

- make every effort to meet the specified 'date required'. If this is not possible the contractor must ensure that he is informed promptly so that additional enquiries can be sent out in order to maintain a full enquiry list

- clear queries as they arise in order to avoid quotations marked 'more information required'

- submit the quotation on time with a clear statement where prices are 'to follow'.

Particular attention should be paid to material supply conditions which may have cost implications. The contractor can incur costs for:

- pallets left on site

- standing time for vehicles while unloading at the site

- small quantities or abnormal loads.

Consideration should also be given and, the cost should be established, for various additional matters associated with materials. These include:

- any specific divergence or discrepancies from the contractor's enquiry in the quotations received from the supplier

- any minimum delivery requirements and adjustment of cost due to delivery in small quantities

- trade discounts, which should be noted separately and reported at the settlement meeting. (Note that discounts may or may not be deducted from the materials cost at this stage. Some contractors maintain that materials costs should be net of discounts, which are summarised in the estimate summaries. Others allow the discount to remain in the materials cost but recognise the element when considering the profit mark-up at the settlement stage.)

- waste allowances

- unloading, storage and distribution costs.

The degree of mechanisation in unloading must be considered, to ensure that material deliveries are compatible with the intended method of handling; such as palleted materials for handling by forklift truck. Special equipment should be considered for unloading, although the costs of skips, slings and chains are more likely to be costed in project overheads than allocated directly against unit rates. The labour costs of unloading and distributing materials must be considered and an allowance made when establishing the total labour requirements of the project. Such labour can either be taken into account when selecting production standards for labour, or can be priced as a project overhead. Items to be accommodated include:

- storage needs and protection

- size and weight of materials

- the cost of any special packaging and crates, if these are charged, or the cost of returning them to the supplier

- any subsidiary fixing materials or temporary materials needed for storage.

The tender works programme and method statement indicate the anticipated:

- time and rate of delivery required for materials

- amount to be stored on site, the location and method of subsequent distribution

- unloading point.

The visit to the site and locality shows:

- physical conditions and any restrictions likely to be encountered

- site layout, operating, storage and unloading facilities.

The allowance made for waste must, wherever possible, be based on experience gained on previous projects. Data given in textbooks, periodicals and manufacturers' catalogues should be examined critically and, used with caution. The waste allowance must be carefully applied according to the circumstances of the project and previous experience of the material. Particular attention should also be given to any implications of the environmental regulations including Site Waste Management Regulations 2008.

The identification and costing of these various factors and considerations will convert the basic cost contained in the quotation into the cost which will be inserted into the net unit rates.

6.7   Sub-contract quotations

Most contracts have provision for the contractor to sub-let work. The analysis of quotations received for domestic sub-contractors is set out in the domestic sub-contractors' register (see Chapter 4). This analysis identifies any further matters which have to be costed by the contractor. Selection of the sub-contractor to be used may not be possible before such additional costs have been determined.

Allowances must now be made for any additional matters associated with the domestic sub-contractor's works. These can include:

■   Specific divergence or discrepancy from the contractor's enquiry included in the quotation.

■   Allowance for unloading, storage and protection of materials and equipment and transfer of goods from stores to point of work, if this is to be the main contractor's responsibility. The labour costs associated with unloading and distribution of materials are considered at this time and allowance made when establishing the total labour requirements of the project. Such labour is either taken into account by an addition to the sub-contractor's quotation or can be priced in the project overheads.

■   General attendance items to be provided by the main contractor.

In making such allowances, the contractor must take into account the requirements of the tender works programme and method statement and facilities which have already been allocated for the contractor's own works. Additions made to cover attendance on domestic sub-contractors may be done in several ways, by:

■   increasing the relevant unit rates of the sub-contracted work

■   adding a fixed percentage to the whole of the sub-contractor's quotation

■   making an addition subsequently in the project overheads.

Discounts offered by sub-contractors must be noted separately and reported at the settlement meeting. (Note: discounts may or may not be deducted from the sub-contractor's quotation at this stage. Some contractors maintain that sub-contractor's costs should be net of discounts, which are summarised in the summary reports. Others allow the discounts to remain in the cost of the work to be sub-contracted but recognise the element when considering the profit mark-up at settlement stage.)

Great care must be taken in assessing sub-contract quotations to ensure that all items have been adequately covered. If labour-only sub-contractors are being considered, the cost allowance *must take into account all factors associated with the provision of materials by the contractor*, and adequate safeguards must be made to control the use and wastage of such materials.

Pricing the works

Figure 6.3 A specimen plant quotations register.

			PLANT QUOTATIONS REGISTER						Project: Helix Laboratories, Westfield		
Ref.	Plant item	Unit	Basic hire rate	Operator cost	Fuels	Maintenance costs	Delivery and collection	Totals	Rates used	Remarks	
	Plant Department										
	JCB3CX	hr	5.00	8.00	1.50	1.50		16.00			
	20T backacter	hr	7.50	8.00	3.00	2.00		20.50	20.50		
	Forklift	week	180.00	inc	30.00	15.00		225.00	225.00	driver in proj o/heads	
	DOLPHIN PLANT										
	JCB3CX	hr	11.50	inc	1.50	1.50		14.50	14.50	fuel not quoted	
	20T backacter	hr	16.50	inc	3.00	2.00		21.50		fuel not quoted	
	Forklift	week								not available	

6.8   Unit rate calculation

Pricing notes

There are many techniques used by estimators in building up unit rates, ranging from rough notes in the bills of quantities, to sophisticated forms showing the constituents of rates broken down into rates and totals for labour, materials, plant and sub-contract elements. If changes are made to the estimate at the settlement meeting, the estimator's handwritten pricing notes could not be used (without further explanation) by construction staff except as a guide to the general logic and pricing structure.

A computer system, on the other hand, will enable an estimator to incorporate all the adjustments made to an estimate, and thus produce a valuable cost control document for site. The strengths of computer-aided estimating systems are the rapid re-calculation and reporting facilities. One drawback is the difficulty management may have in gaining an understanding of how rates were built up.

Unit rate checklist

A unit rate checklist can be used to remind the estimator of the main constituents of a unit rate. The example given in Fig. 6.3 distinguishes between costs which are incorporated in the rate, and those which could be priced in the project overheads.

For clarity, Fig. 6.4 does not include a sub-contract element which could either be the whole rate or one of the constituents. Sub-contract rates are commonly added to rates for alterations and repairs such as 'remove floor coverings from main entrance and apply levelling compound'. In this example, the levelling compound work may be part of a flooring specialist's package.

Pricing the works

Pricing the works

Figure 6.4 A specimen schedule of PC sums and attendances.

SCHEDULE OF PC SUMS AND ATTENDANCES

Project: Helix Laboratories, Westfield

PC sums

Bill page	item	Nominated suppliers	gross	discount	net	Attendances			
						lab	pit	mat	sub
5/3	a	Ironmongery	5000	250	4750				
	b	Doors	11000	550	10450				
	c	Kitchen appliances	3500	175	3325				
		Totals to summary	19500	975	18525				

Bill page	item	Nominated sub-contractors	gross	discount	net				
5/4	a	Structural steelwork	45000	1125	43875				
	b	Lift installation	24000	600	23400				
	c	Aluminium windows	74000	1850	72150				
	g	Access road for steelwork				300	300	600	
		Totals to summary	143000	3575	139425	300	300	600	

6.9   PC and provisional sums

Figure 6.5 PC and provisional sums flowchart.

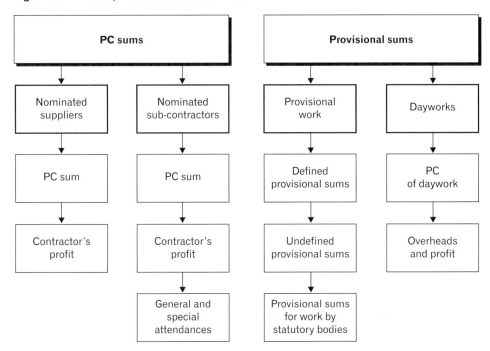

Different contract forms have differing provisions regarding the naming or nominating of suppliers and sub-contractors. The JCT 2005 Standard Building Contract does not have provision for nominations, although these could be included by having a supplemental agreement. As naming or nomination suppliers may still be used on some contracts, guidance on these is included here.

Named/nominated suppliers

Standard methods of measurement state that the cost of materials from nominated suppliers is to be identified in the tender documents as PC sums. A separate item is also given for the contractor to add his profit. PC sums may also be written in to an item description (such as a rate for the supply of facing bricks) for the estimator to incorporate the cost in his rate build-up. Named suppliers effectively become a domestic supplier and would not be identified separately in the tender documentation for costing purposes.

The estimator will produce a list of nominated suppliers at an early stage using the schedule of PC sums and attendances (Fig. 6.4). Where details associated with a nominated supplier are unclear, the estimator must note any concerns in his report for further consideration at the settlement meeting.

If a PC sum has been included for high-value materials or large quantities, the estimator must check the following:

- The terms of the purchase contract – the estimator should check if any discounts have been allowed by the suppliers. This discount is normally deducted from the estimate in order to include net costs in the summaries for the settlement meeting.

- Delivery times and how they affect the tender works programme.

- Fixing items associated with materials provided by a nominated supplier need to be described adequately and measured in the items to be priced. Any discrepancies concerning fixings, such as bolts, screws, brackets, adhesives and sealants, or ambiguity over the responsibility for supply of these items must be clarified.

■ Additional costs for unpacking, storage, handling, hoisting and the return of reusable crates or pallets to the supplier. Suppliers may deliver their materials in reusable crates or other packaging and the contractor may be required to return such items to the supplier. Due allowance must be made for the collection, storage, handling and subsequent dispatch of such items back to the supplier.

Where bills of quantities are used, the fixing of materials supplied by nominated suppliers is measured in the appropriate part of the bill.

Named/nominated sub-contractors

A nomination arises in construction contracts where the selection of a sub-contractor is to be made by the client or his representative, for which a PC sum has been inserted in the tender documents. Again, the contract should be checked to ensure that this is permissible and that supplements have been provided if no main contract clauses exist.

There is a right of reasonable objection to a particular nominated sub-contractor because it would be contrary to contract law for a party to be forced into a contract against their will where they had reasonable cause for not wishing to contract with a particular nominated party. Where PC sums are included in bills of quantities, the estimator is seldom given the name of the proposed sub-contractor and so it is rarely possible to discuss methods and project planning issues prior to tender.

The use of nominated sub-contractors is becoming increasingly rare for two reasons:

■ the growing complexity of nomination procedures

■ the added risks carried by a client when a contractor is relieved from a proportion of the responsibility for full performance.

Where nominations are used they can account for a significant proportion of the overall cost of a contract. Contractors are frequently given inadequate supporting information to deal with attendances in the bills of quantities. The adequacy of the information provided must be carefully investigated, and further particulars requested by the estimator if details are not complete. This enables the estimator to be able to include nomination dates, nominal design, fabrication, delivery work period dates and interfaces for work in the tender works programme.

Adequacy of information

The standard method of measurement gives the items to be included in the bills of quantities for each nominated sub-contractor, as follows:

■ the nature and construction of the work

■ a statement of how and where the work is to be fixed

■ quantities which indicate the scope of the work

■ any employers' limitations affecting the method or timing, sequence or sectional completion of the works

■ a PC sum

- general attendance item

- an item for main contractor's profit, to be shown as a percentage

- details of special attendance required by the sub-contractor.

At tender stage, the estimator must check that the measured items for works which are covered by a PC sum are adequate and that supporting details are available in accordance with the appropriate Standard Method of Measurement. If not, it will be necessary to ask for further particulars before the inter-relationships of trades, the tender works programme and method statement can be finalised. All too often this is a neglected area and the estimator must ensure that the tender stage construction programme provides sufficient time for the execution of work which is the subject of a PC sum and that all trades are effectively co-ordinated.

If there are any doubts as to the adequacy or meaning of the descriptions used, clarification must be obtained from consultants.

Attendances

Attendance is defined as, *'the labour, plant, materials or other facilities provided by the main contractor for the benefit of the sub-contractor and for which the sub-contractor normally bears no cost'.* The main contractor is responsible under the main contract provisions for the site establishment and providing attendance. This provides clear responsibilities for the support services and equipment needed on site and eliminates duplication of resources for various specialist sub-contractors. For very large contracts, where a construction manager or management contractor has overall control, trade contractors are asked to provide certain parts of the temporary works and facilities themselves. The costs associated with attendance are built into the main contractor's tender and consequently become a charge against the client. However, the associated risks of attendance are borne by the main contractor.

The estimator must decide how to price 'general attendance' and 'special attendance' relating to nominated sub-contractors. The attendances may be priced in the project overheads schedules or on the schedule of PC sums and attendances (Fig. 6.4).

General attendance

The item for general attendance is an indication of the facilities which are normally available to sub-contractors and that they are provided by the contractor to meet regulatory and logistical requirements.

In assessing any sums to be allowed for general attendance, the estimator must investigate the facilities which will already be provided for the main contractor's use and determine any costs which may arise by the nominated sub-contractor's use of any such facilities. The facilities given in SMM7 include:

- use of temporary roads, paving and paths

- use of standing scaffolding

- use of standing power-operated hoisting plant

- use of mess rooms, sanitary accommodation and welfare facilities

- provision of temporary lighting and water supplies

- providing space for sub-contractor's own office accommodation and for storage of his plant and materials

- clearing away rubbish.

Use of temporary roads, paving and slabs

Allowance must be made for any costs associated with the maintenance of temporary roads, paving and paths which are required during the time period allowed by the contractor for his own use. This item will not cover any specific access requirements of a nominated sub-contractor. Such items, for example, as hard standing for a crane should be separately described under 'special attendance'.

Use of standing scaffolding

The contractor must allow for any costs which might arise through the nominated sub-contractor's use of scaffolding which is already erected for the main contractor's use. Any modifications or additional scaffolding required, or, any increase in duration for such scaffolding over and above the time period required by the main contractor, must be described and measured as 'special attendance' in the bills of quantities.

Use of standing power-operated hoisting plant

While nominated sub-contractors may use existing hoisting plant if there is spare capacity, any hoisting facilities specifically required must be measured under 'special attendance'.

Use of mess rooms, toilets and welfare facilities

Assessment must be made of the accommodation needed for the operations of nominated sub-contractors over and above the requirements of the contractor. Allowance must also be made for any servicing and cleaning of such facilities which are shared with the contractor.

Provision of temporary lighting and water supplies

The estimator must establish requirements for general lighting needed to comply with safety requirements and for the execution of the works during normal working hours. Adequate allowance must also be made for water points needed for the construction of the works. This may mean the simultaneous provision of such services in other areas of the building over and above the requirements of the main contractor.

Special lighting requirements and power needs must be measured and priced under 'special attendance'. Specific water requirements for testing or associated with commissioning of plant should be measured under 'special attendance'.

Providing for sub-contractor's own space

The estimator should note that only space is required and that cover in the form of a shed is not a requirement. The assessment of total space requirements must be borne in mind when finalising the method statement and site layout.

Clearing away rubbish

The disposal of waste, packaging and other rubbish from an agreed collection point involving labour, containers and haulage must be assessed. Abnormal items of rubbish, such as disposing of surplus excavated material from a ground improvement technique, must be measured separately under 'special attendance'. The prevailing regulatory requirements should also be taken into consideration (for example regarding disposal of hazardous waste material).

Special attendance

Other specific attendances which do not fall under the category of 'general attendance' must be specifically measured in the bill of quantities as 'special attendance'. Items to be measured include:

- special scaffolding or scaffolding additional to the contractor's standing scaffolding

- the provision of temporary access roads and hardstandings in connection with structural steelwork, precast concrete components, piling, heavy items of plant

- unloading, distributing, hoisting and placing in position, giving in the case of significant items the weight, location and size

- the provision of covered storage and accommodation, including lighting and power thereto

- power supplies giving the maximum load

- any other attendance not included in 'general attendance' or listed above.

Special scaffolding

In order to price this item, the estimator must be given precise details concerning the scaffolding requirements. Such information should define clearly the height in stages of the scaffolding, indicate the extent of boarded platforms and any alteration and adaptation that will be required. If such information is not available and descriptions are inadequate, the estimator should seek further instructions from the consultant.

The estimator must also make due allowance under this heading for any adaption or alteration to standing scaffolding or for any extension to the time period, providing such items are described and measured in the bills of quantities.

The provision of temporary access roads and hardstandings

Where any specific requirements are described, these must be taken into account with the contractor's own needs, and any additional temporary provision allowed for.

Unloading, distributing, hoisting and placing in position

This item may also include some intermediate storage requirement and due allowance should be made for this. It is essential that particulars are stated of the size or weight of materials to be handled to enable the estimator to reasonably assess costs and identify the appropriate mechanical aids. In the case of heavy units, e.g. precast cladding, it will also be necessary to be informed of the delivery rate and also if any specific stacking facilities are required for storage on site.

Sufficient information must also be provided to identify any distribution requirements, as opposed to hoisting and stacking. In assessing the cost of such specific facilities, the estimator must bear in mind the extent of use of existing mechanical hoists and must ensure that sufficient hoisting capacity is available to meet all needs.

In some cases, the estimator needs to seek clarification if components such as precast concrete cladding panels or mechanical plant are to be placed in position. This is because there could be overlapping responsibilities with the sub-contractor who would be expected to supply and fix all of his own materials.

Provision of covered storage and accommodation including lighting and power

Under 'general attendance' the contractor is required to provide space for nominated sub-contractors to erect their own facilities. Under this item the main contractor will be required to provide, erect and maintain accommodation and provide lighting and power as stipulated. The size of hutting required should be stipulated and the period required stated. Any special requirements, i.e. racking or other services, should also be defined.

Power supplies giving the maximum load

Any special power requirements, including power for testing of systems, must be clearly measured for pricing purposes. Any reference to power supplies should state whether single- or three-phase electrical-power supplies are required, and the maximum demand level should be taken into account. The estimator should ensure that any descriptions for fuel or power for such testing purposes are clearly specified, giving the quantity necessary to fulfil the tests and also the precise specification of the power needs.

Maintenance of specific temperature or humidity levels

Any specific requirements for controlling temperature or humidity must be clearly measured, stating temperature/humidity required and the time period that the contractor must provide these services. The requirement must also state if the permanent services in the building can be used for this purpose.

Any other attendances

Under this heading, the contractor is required to provide specific attendance or materials for various trades. This could include the provision of bedding material for roof tiles, or floor tiles. Other items such as specific cleaning operations and the removal of masking tape used by sub-contractors should also be defined. Such items must be clearly measured and in the event of any inadequacies or ambiguities, the estimator should refer to the consultants for further instructions.

With the exception of provisional sums for work by statutory authorities, provisional sums are deemed to include an allowance for main contractor's head office overheads and profit. All provisional sums will become the subject of an architect's instruction during construction and the work will be valued according to the appropriate contract rules for measurement and valuation which include provision for overheads and profit. Accordingly, the value of provisional sums (except those for statutory authorities) should be added into the final summary after the application of overheads and profit. Alternatively, if it is company practice to add head office overheads and profit to the total value of measured and unmeasured work, provisional sums should be discounted before entry in the summary to avoid duplicating the overheads and profit for this type of work.

Figure 6.6 illustrates a typical schedule of provisional sums and daywork. The total for each category is carried forward to the final summary for review by management.

Provisional sums

Provisional sums are included in bills of quantities for items of work which cannot be fully described or measured in accordance with the rules of the method of measurement at the time of tender.

For work measured under the rules of SMM7, there are three types of provisional sum as listed below.

Provisional sum for defined works

This provisional sum is used where works are known to be required in the project but have not been fully designed or specified at tender stage, and so cannot be measured in detail. The contractor must make due allowance for the planning engineering, project scheduling and pricing preliminaries; and to enable him to do so the following information must be provided with the provisional sum:

- the nature of the work

- how and where it is to be fixed

- quantities showing the scope and extent of the work

- limitations on method, sequence and timing.

Example of provisional sum for defined work: 'Allow the provisional sum of £8000 for the reception counter to be located in zone G3. The counter will be L-shaped on plan, approximately 6.00 m girth and incorporate four workstations, and is to be installed after completion of floor finishes.'

Pricing the works

Figure 6.6 A typical schedule of provisional sums and daywork.

Project:		SCHEDULE OF PROVISIONAL SUMS AND DAYWORKS
Helix Laboratories, Westfield		

PROVISIONAL SUMS

Bill page	item	Defined provisional sums	net
5/6	a	Security fencing to rear yard	5000
	b	Reception desk	11000
	c	Kitchen appliances	3500
		Totals to summary	£ 19500

Bill page	item	Undefined provisional sums	gross	mark-up*	net
5/8	a	Entrance gates	11000	(1100)	9900
	b	Additional insulation to roof space	2500	(250)	2250
	c	Machine bases in plant room	4500	(450)	4050
	d	Additional builder's work	5000	(500)	4500
5/9	a–g	Various	22000	(2200)	19800
5/11	d	Contingencies	40000	(4000)	36000
		Total to summary	85000	(8500)	76500

* Possible over-recovery of overheads and profit

DAYWORKS

Bill page	item	type	basic sum	add	deduct	total
5/12	a	Building labour	8000	8000		16000
	c	Specialist labour	8000	9600		17600
		Labour totals	16000	17600	0	33600
5/13	a	Building materials	8000	1200		9200
	c	Specialist materials	8000	1200		9200
		Material totals	16000	2400	0	18400
5/14	a	Building plant	4000	400		4400
	b	Specialist plant	4000	400		4400
		Plant totals	8000	800	0	8800
		Daywork totals	40000	20800	0	60800

Provisional sum for undefined works

Where the information required in support of a defined provisional sum is not available, the provisional sum is 'undefined' and the contractor is not required to include any duration for the work in the tender works programme, nor to take account of the costs of planning engineering, project scheduling or preliminaries.

Undefined provisional sums are typically used to make contingent provision for possible expenditure on elements of work which cannot be wholly foreseen at tender stage or cannot be quantified. A client's contingency sum is deemed to be an undefined provisional sum.

Example of provisional sum for undefined work: 'Allow the provisional sum of £750 for remedial work to roof boarding exposed, after the removal of existing roof coverings.'

If the work required to the roof is subsequently instructed, this would become a variation to the contract and valued according to the main contract provisions.

Provisional sum for works by statutory authorities

SMM7 makes provision for a provisional sum to be included in a bill of quantities which is neither 'defined' nor 'undefined', for work to be carried out by the local authority or statutory undertakings, including privatised services authorities carrying out statutory works.

Incorporating provisional sums in an estimate

Although not strictly required by SMM7, it is common for items to be measured to enable the contractor to price attendance and profit on works by statutory authorities as though they were nominated sub-contractors. It should be noted that these provisional sums are net and do not include a contractor's discount.

Daywork

Definition

Contractors must understand the circumstances in which varied or additional work will be valued on a daywork basis. It normally occurs where variations cannot be valued by measurement using bill rates or comparable rates, nor by negotiation before an instruction is issued.

The daywork charges are usually calculated using the definitions for prime costs and overheads published by the RICS/BEC for building work and FCEC for civil engineering.

The prime cost of daywork can be defined in other ways, so care must be exercised in reading the definition in the tender documents.

The composition of the total daywork charge will include the following costs:

- labour

- materials and goods

- plant

- supplementary charges (civil engineering contracts)

- incidental costs, overheads and profit (this addition will vary between labour, plant and materials and, in order to introduce competition at tender stage, is added to provisional sums for the prime cost of labour in the bills of quantities by the contractor).

An alternative method (for labour to be valued on a daywork basis) is for the contractor to provide all-in gross hourly rates which are applied to provisional hours. This makes the calculation of daywork rates simpler during the course of the project, but moves the burden for anticipating increased costs to the contractor.

Contractors may decide that some of the project and head office overheads are covered in the contract price and may be excluded from daywork rates. This is mainly true if the daywork, to be carried out during the currency of the contract, will not result in an extension to the contract, but other additional costs to project and head office overheads may still have to be considered.

It is inappropriate to use this payment method for anything except work which is incidental to contract work. In the event that significant changes are made to the original scope of works the valuation rules normally allow additional overhead costs to be recovered, usually when the full effects of changes are known.

Decisions concerning allowances for profit and overheads must be made by each contractor taking into account his own circumstances and method of working and his assessment of the effects of daywork on a particular project. The contractor must assess each contract on its own merits in producing daywork rates and calculating the percentage addition needed. This will include an assessment of the likelihood of the prime cost being a reasonable pre-estimate of the work, which will be valued on a daywork basis.

The contractor's daywork percentages must take into account the rates required by the sub-contractors used in the tender. Enquiries to sub-contractors must include a request for daywork percentages based on the definition incorporated in the main contract. For mechanical and electrical installations in building contracts, the contractor is given the facility to state different percentages for specialists in the bills of quantities.

As dayworks are calculated inclusive of an allowance for overheads and profit, they should, like provisional sums, be added into the final summary after the application of overheads and profit.

Labour
The calculation for the PC of labour differs somewhat from that of the all-in rate; various incidental costs, overheads and profit are deemed to be included in the percentage added to the PC by the contractor. The percentage added must include all other items that the contractor considers are necessary to recover the true cost incurred but are not allowed under the definition of PC of daywork.

The PC of labour calculation consists of:

- guaranteed minimum wages and emoluments

- additional emoluments in respect of the Working Rule Agreement

- overhead costs in employing operatives

- incidental costs, overheads and profit: the percentage addition.

To calculate the percentage to add to the PC for daywork for labour, the estimator must find the difference between the total cost of labour and the PC calculated using the definition.

Figure 6.7 uses the RICS/BEC definition of PC of daywork to compare the PC calculation for labour with the all-in rate produced in Section 6.3. The calculation shows that in this example £2.86 (51%) has to be added to the PC of labour for incidental costs, overheads and profit. This is the percentage to be inserted in the bills of quantities.

In practice, contractors often add considerably higher percentages than this for the following reasons:

- to provide for additional supervision and record keeping

Figure 6.7 Example. Calculation of daywork labour rates.

Description	Quantity	Rate	Prime cost of labour for daywork £	Items not included in definition of prime cost
(a) Annual prime cost of labour				
Guaranteed minimum wage	1870	4.58	8564.60	
Extra payment for skill				
Public holidays	8	35.72	285.79	
Employer's national insurance contribution	8565	7%	599.55	
Contribution to annual holiday credits	47	17.55	824.85	
Contributions to death benefit scheme	47	2.05	96.35	
Contributions, levy, tax payable by employer	8565	0.25%	21.41	
Annual total for labour			10,392.55	
Divided by 1870 hours for hourly rate			5.56	
(b) Incidental costs overheads and profit				☐
Head office charges				☐
Site staff				☐
Trade supervision not working manually				☐
Additional cost of overtime				☐
Time lost due to inclement weather				☐
Additional bonuses and incentive schemes				☐
Subsistence and periodic allowances				☐
Fares and travelling allowances				☐
Sick pay				☐
Third party and employer's liability insurance				☐
Liability in respect of redundancy payments				☐
Tool allowances				☐
Use, repair and sharpening of non-mech tools				☐
Use of erected scaffolding, staging, trestles				☐
Use of tarpaulins, protective clothing, artificial lighting, safety and welfare, etc.				☐
Profit				☐

Calculation of percentage required for incidental costs

All-in hourly rate = 7.14
    add site staff costs (7%)
    add head office charges (5%)
    add profit (5%)

All-in rate required for labour = 8.42
    less prime cost (given in table) = 5.56

Addition required = 8.42 − 5.56 = 2.86

Percentage addition required for incidental costs = **2.86/5.56 = 51%**

- to account for the disruption to the construction programme

- for the use of more costly labour-only specialist trade sub-contractors

- the type of work associated with daywork expenditure may require a higher level of supervision.

Builder's work Items of builder's work in connection with works by nominated sub-contractors will be measured in accordance with the requirements of the Standard Method of Measurement. The estimator will then be required to establish unit rates for the measured items in accordance with the principles previously described. The following points should be borne in mind:

- The extent to which builder's work items are shown on the tender drawings. It may be necessary to ask for further particulars before pricing can be completed.

- The requirement for any specialist work or plant to carry out the builder's work required.

- The rate and timing at which the builder's work should be carried out and the continuity to be expected whilst carrying out this builder's work.

With daywork charges included in a competitive tender, the contractor must look for ways to minimise costs. If the provisional sum for dayworks is large then contractors may insert lower percentage additions. Conversely where insufficient sums are allowed for daywork, contractors may add greater percentages and assess the possible over-recovery of margin in the settlement meeting.

Materials and goods The PC of materials and goods obtained from stockists or manufacturers is the invoice cost after deduction of all trade discounts but, including cash discounts not exceeding 5% and, including the cost of delivery to site. Since the material cost is that which is invoiced by the supplier, it may be assumed that the prime cost includes overlaps and waste. Similarly, the costs of packing materials, handling unloading, storage and returning packing cases are also part of the prime cost allowance.

The PC of materials and goods supplied from the contractor's stock is based on the current market prices, plus any appropriate handling charges.

The percentage addition will be based on the contractor's allowance needed for site and head office overheads and profit. Typical amounts quoted are 10% for site overheads, 5% for head office overheads and 5% for profit. For a particular project, an estimator may take the view that some of these costs will not be incurred and so can be ignored for the purpose of winning the contract.

Plant The RICS/BEC definition includes a Schedule of Basic Plant Charges. The rates in the schedule are intended to apply solely to daywork carried out under and, incidental to, a building contract. They are *not* intended to apply to:

- jobbing or any other work carried out as a main or separate contract; or

- work carried out after the date of commencement of the defects liability period.

The rates apply to plant and machinery already on site, whether hired or owned by the contractor. Unless otherwise stated, the rates include the cost of fuel of every description; lubricating oils, grease, maintenance, sharpening of tools, replacement of spare parts, all consumable stores and for licences and insurances applicable to items of plant. They do not include the costs of drivers and attendants.

The rates must be applied to the time during which the plant is actually engaged in daywork. Whether or not plant is chargeable on daywork depends on the daywork agreement in use. The inclusion of an item of plant in the schedule does not necessarily indicate that the item is in fact chargeable.

Rates for plant not included in the schedule* or which are not on site, and are specifically hired for daywork, shall be settled at prices which are reasonably related to the rates in the schedule, having regard to any overall adjustment quoted by the contractor in the conditions of contract.

The contractor will need to allow for various additions to the basic plant rates when compiling the daywork rate for plant. These include:

■   general overheads, incidental costs and profit

■   the effects of inflation since the publication date of the Schedule of Basic Plant Charges

■   the differences between rates given in the schedule and the anticipated rates actually being paid for plant.

6.10   Preliminaries and project overheads

Figure 6.8 Preliminaries and project overheads.

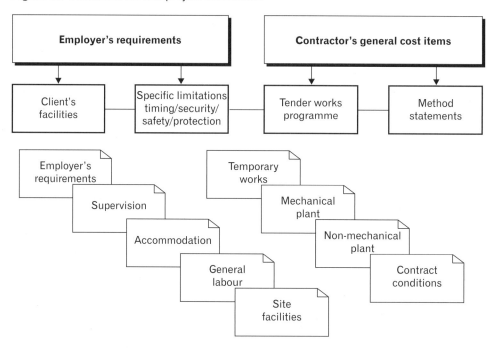

 The standard methods of measurement for civil engineering and building give the general items which should be described in a bill of quantities, in two main parts: the specific requirements of the employer and the facilities which must be provided by the contractor to carry out the work. Items listed in the preliminaries/general conditions section of a bill of quantities are not exhaustive and are provided to produce a framework for the convenience of pricing.

The actual supervision, site facilities, plant and temporary works to be employed will be decided by the contractor unless it is clearly stated that certain systems

---

* Items such as small plant and loose tools, non-mechanical based tools, erected scaffolding, staging, trestles and the like are excluded from the schedule of plant items. They, together with consumable stores and protective clothing, are normally incorporated into the percentage addition added to the labour rate.

must be used. A temporary roof may be required above an historical building, for example, if the employer is convinced that such an arrangement is the only way to protect the valuable contents of a building during a re-roofing contract. On the other hand the decision to employ a temporary roof may be left to the tenderers.

Figure 6.9 shows a checklist used by an estimator to list the items for inclusion in the project overheads. The points come from the following sources:

- a careful reading of the preliminaries bill

- various items noted during the pricing stage

- the preparation of and updating of the method statements

- the preparation of and updating of the critical path network master works programme

- the preparation of and integration of resourced short-term critical path network works programmes

- record keeping.

Figure 6.9 A specimen project overheads checklist.

	PROJECT OVERHEADS CHECKLIST	Tender number 97/201

Ref.	Description	Section	Complete
1.3d	Bond required 10% for 24 months	Contract conditions	✓
1.4c	Principal contractor – prepare safety plan	Miscellaneous	✓
1.7f	Liquidated damages £5500 per week	Contract conditions	✓
1.12d	Office for clerk of works 15 m² plus furniture	Empl requirements	✓
1.12f	Telephone for clerk of works – install and charges	Empl requirements	✓
1.15a	Co-ordination of services – qualified engineer	Supervision	✓
1.17g	Fax machine required on site	Site facilities	✓
1.18f	Rubbish to be removed daily	Labour/plant	✓
1.18h	Hoarding required to south boundary – drg 406	Temp works	✓
1.18k	Notice board	Temp works	✓
1.20c	Protection for conc columns	Temp works	✓
1.24b	Final clean check phased completion	General labour	✓
	ATTENDANCES		
S14	Check additional scaffolding for ceilings	Non-mech plant	✓
S16	Check mixer for ceramic tiling	Mech plant	not required
	GENERALLY		
	Check concrete skip for crane	Mech plant	✓
	Check additional disposal for piling – 20 skips	Non-mech plant	✓

SMM7 defines a *fixed charge* as the cost of work which is to be considered as independent of duration, and a *time-related charge* as the cost of work which is to be considered as dependent on duration. Items in bills of quantities for employer's requirements and contractor's general cost items are given for both categories. The purpose is to improve the allocation of values in valuations for interim payments and variations. Some bills of quantities list *value-related items* for general items such as insurances, bonds and the supply of water to the site, on the bill summary page.

Where the procurement route chosen is design and build then a different approach to estimating is used. Here the 'top-down' method is more appropriate as the client will normally have established a budget and be looking for a design that can be achieved with this. This is explained in Chapter 11.

Appendix

A6.1 **Calculation of hours worked**

Annual and public holidays

	Annual holidays (21 days)	**Public holidays (8 days)**
Winter	7 working days at Christmas	Christmas day Boxing day New Year's day
Summer	4 working days at Easter 10 working days (2 calendar weeks) in summer	Good Friday Easter Monday First Monday in May Spring Late summer

Calculation of hours worked

Summer period	*Hours*	*Winter period*	*Hours*
starting time 8.00am		starting time 8.00am	
finishing time 5.30pm		finishing time 4.30pm	
Friday 4.30pm		Friday 3.30pm	
lunch period 1.00–1.30pm		lunch period 1.00–1.30pm	
working week = 44 hrs		working week = 39 hrs	
for 30 weeks =	1320	for 22 weeks =	858
deduct:		deduct:	
10 days summer holiday	(88)	7 days Christmas holiday	(55)
4 days Easter holiday	(35)	3 days public holiday	(23)
5 days public holiday	(44)		
Total hours for summer	1153	Total hours for winter	780

Gross hours available for work	1153 + 780 = 1933
Deduct allowance for sickness	
say 8 days in winter	(62)
Net hours available for work (basic hours)	1871

Inclement weather

The time lost for inclement weather will vary according to the type of work, season of the year and geographical area. An average allowance is used and any adjust-

Pricing the works

ment necessary made for exceptional situations in the project overheads. Say the time lost due to inclement weather is 2%, i.e. approximately 37 hours.

Actual hours worked = basic *less* inclement weather   1871 – 37 = 1834.

A6.2   Calculation of annual employment costs

Guaranteed minimum wages and emoluments (annual costs)

Note: All figures quoted are those *current* as at 27 June 1996.

(a) Guaranteed minimum wages

Hours worked per year (basic hours)	1871

With weekly rate of pay at 27 June 1996

Guaranteed minimum weekly earnings (basic rate)

is £178.62 per 39 hour week

Basic wages are: 1871hrs x £4.58/hr =   £8569.18

(b) Contractor's bonus allowance

Say £20 per 39 hour week   1871 hrs x £0.51/hr =   £954.21

(c) Inclement weather allowance

This is included in guaranteed minimum wage above.

(d) Non-productive overtime costs

Overtime rates are paid for time worked in excess of the normal working hours.

This allowance is for overtime worked as normal practice; the cost of any additional special overtime will be dealt with separately in the project overheads.

Overtime

The overtime allowance is calculated as follows:

■   Summer period:

	Mon.	Tue.	Wed.	Thu.	Fri.	Sat.	Sun.	Total
Hours worked	9	9	9	9	8			44
Basic hours	8	8	8	8	7			39
Overtime hours								
	1	1	1	1	1			5
Non-productive hours	"	"	"	"	"			2"

Working weeks in summer 30 less 4 weeks for annual and public holidays = 26.

Non-productive overtime for summer period = 26 weeks x 2.5 hrs = 65 hrs.

- Winter period

	Mon.	Tue.	Wed.	Thu.	Fri.	Sat.	Sun.	Total
Hours worked	8	8	8	8	7			39
Basic hours	8	8	8	8	7			39
Overtime hours	0	0	0	0	0			0
Non-productive hours	0	0	0	0	0			0

Total for whole year 65 hrs. Therefore, cost of non-productive overtime: (basic rate per 39-hour week = £178.62/39 = £4.58). Therefore overtime allowance = £4.58 x 65 = £297.70.

Sick pay

For 8 days of sickness the first 3 qualifying days are not due for payment. Assumed number of payable days lost due to sickness is 5. Statutory rate of payment is £12.10 per day. Therefore, cost of sick pay is 5 x £12.10 = £60.50

If a private insurance scheme is used instead of a statutory scheme, the cost of premiums should be included here instead of the statutory allowances.

Trade supervision

The number of trades foremen to operatives will vary from company to company and in accordance with the needs of a project. Assume that:

- there is one trades foreman for every eight tradesmen

- half their time is spent working and half on supervisory duties

- their rate of pay is £0.32 per hour above the trade rate.

Hourly cost for the gang is

1 trades foreman	£4.58 + £0.32	=	£4.90
8 tradesmen	£4.58	=	£36.64
		=	£41.54

Allowing for supervision time, the effective hourly rate for working operative is:

$$£41.54 / 8.5 = £4.89$$

Therefore, additional hourly cost of supervision is:

$$£4.89 - £4.58 = £0.31/hr$$

Hours worked per year = 1871.

Therefore, cost of trade supervision is:

$$1871 \times £0.31 = £580.01$$

It is normal to use a firm's individual arrangements in basic calculations and adjust in the project overheads (if necessary for a particular project).

Pricing the works

Working Rule Agreement allowances

Operatives exercising special skills or working in particular circumstances, are entitled to special allowances under the Working Rule Agreement. Examples include discomfort, inconvenience or risk, continuous extra skills or responsibility, intermittent responsibility, tool allowances, special provisions or servicing of mechanical plant, storage of tools and clothing. The amount allowed is, therefore, variable according to responsibility and skills. The following is used as an example for the driver of a rough terrain forklift truck up to 3000 kg capacity.

The allowance (for Working Rule Agreement Allowances) to be paid is £14.46 per week or £0.37 per hour.

Therefore, cost of WRA allowance is:

$$1871 \text{ hrs} \times £0.37 \qquad = \qquad £692.27$$

Since the extra payments form part of an operative's basic rate, they must be included in the overtime calculation, as follows:

$$65 \text{ hrs overtime} \times £0.37 \qquad = \qquad £24.05$$

The additional payments are mainly supplements for general operatives and not craft operatives.

Overheads in employing labour

Training

Many companies used to have their own training departments and the costs were normally carried in head office overheads. Training is still undertaken, whether at the company's cost or sponsored by Construction Industry Training Board (CITB) schemes. The operating costs of the CITB are recovered by way of a levy (0.25%) of annual payroll rates applied to managerial, clerical and operative employees.

Where labour-only services are used, the levy is based on a percentage of payments made for labour-only at a rate of 2%. Firms whose payrolls, together with payments for labour-only services are less than a lower limit during the year, are excluded from payment of the levy.

Each company will vary in the amount actually paid to the CITB but for directly employed labour, the 0.25% payment would apply.

National Insurance contributions (not contracted out)

There are different levels of contributions depending on the employee's gross weekly earnings. From 6 April 1996, for weekly earnings in the range £110.00 – £154.99 the employer's contribution is 5% on whole earnings. Since most operatives' earnings will fall into the next range £155.00 – £209.99, the higher rate of 7% applies. (It should be noted that National Insurance contributions are payable on all payments, including productivity bonus which may not be included in the all-in labour rate. It may therefore be necessary to make an assessment of the anticipated value of productivity bonus payments and apply the appropriate percentage National Insurance to the total earnings value.)

Holiday credits (including death benefit scheme)

Employers make weekly payments on behalf of employees for annual holidays and the cost is the same for tradesmen and general operatives. The cost of annual holiday stamp is £19.60 per week (including a £2.05 contribution towards the death benefit scheme). Stamps do not have to be paid during holiday weeks therefore the cost per year = 47 × £19.60 = £921.20.

The cost of public holidays must be met by the employer. Paid public holidays must be allowed by the estimator based on the guaranteed minimum wage for

Pricing the works

each grade of employee for the eight public holidays each year. The cost per year is therefore = £198.62/5 days x 8 (days) = £317.79.

Tool allowances

Although tool allowances are not part of an operative's basic wage, they are subject to National Insurance contributions and the deduction of tax. The tool allowances given in WRA 18 are not taken into account in the calculation of overtime. Carpenters, for example, are entitled to £1.94 per week for the provision, maintenance and upkeep of tools which they provide. Most other trades receive £0.99 per week.

Severance pay and sundry costs

For the following indeterminate factors allow 2% on labour costs:

■   severance pay

■   loss of production during notice

■   absenteeism (the cost of National Insurance, pensions and holidays with pay being spread over a smaller number of working hours than the normal conditions assumed)

■   abortive insurances (paying stamps for operatives who work on Monday in order to have cards stamped but who are absent subsequently).

The percentage allowed will vary from firm to firm according to experience.

Employer's liability and third-party insurance

For employer's liability and third-party insurance, allow 2% on labour. This percentage will vary according to the firm, the insurance company, the company's insurance record, and type of work and size of the contract. It is becoming increasingly common for companies to express all their insurance costs as a percentage of turnover (or in the case of any particular tender, tender price) and it may be appropriate to include this calculation in the project overheads schedule when the full value of the project is known.

A6.3   Summary

Table A6.1 Craft operative 1996–97

a.	Guaranteed minimum wages	8569
b.	Contractor's bonus allowance	954
c.	Inclement weather allowance	Incl.
d.	Non-productive overtime costs	298
e.	Sick pay allowance	60
f.	Trade supervision	580
g.	Working Rule Agreement allowances (generally for labourers not craft operatives)	na
	Sub-total	**10,461**
h.	CITB training contributions (0.25% on annual payroll £10,461)	26
i.	National Insurance contributions (7% of average earnings £200 × 52wks)	728
j.	Holiday credits	1239
k.	Tool allowances	100
	Sub-total	**12,554**
l.	Severance payments (2% of labour costs)	251
	Sub-total	**12,805**
m.	Employer's liability insurance (2% of labour costs)	256
n.	Annual cost per craft operative	13,061
o.	Number of productive hours worked	1834
	All-in hourly rate (n ÷ o)	**£7.12**

Pricing the works

Table A6.1 shows a typical all-in rate calculation produced using spreadsheet software. Estimators use this approach for rapid changes to certain variables such as bonuses and the number of hours worked each week. There are similar facilities in estimating packages which link this calculation to labour resources and remain linked to the build-up of an estimate.

7 Completion of estimate

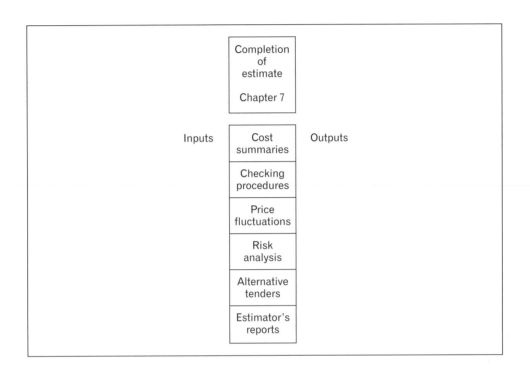

7.1 Cost summaries

An estimator must prepare summaries of the resources making up an estimate so that management can assess the sources of labour, plant, materials and services, their price levels, price comparisons between suppliers and possible discounts available. The forms given in Figs 7.1–7.4 illustrate how the constituents of an estimate are summarised clearly in a way which helps management to make decisions using accurate information.

7.2 Checking procedures

Once the items and quantities have been entered into the computer correctly, the estimator is normally responsible for accurately building up unit rates; his work is checked in four ways:

1. Forms can be designed to add across and downwards so that a page total comes from two directions. Examples of self-checking forms can be seen in the project overheads summaries.

2. By questioning during the review meeting, an estimator will explain how the principal rates were calculated.

Completion of estimate

Figure 7.1 A specimen labour summary form.

LABOUR SUMMARY

Project: Helix Laboratories, Westfield

Ref.	Description	Trade	Total hrs	All-in rate £/hr	Totals in rates £	Typical outputs	Labour-only			Late quotations			Adjusted totals
							Name	Quote	Adjustment	Name	Quote	Adjustment	
L01	Concrete work	Conc labourer	1200	6.00	7200	frids 2.0 hr/m^3, slab 1.6 hr/m^3	MK Formwork	6900	−300	F. Smith	6653	−247	6653
L02		Carpenter	300	9.00	2700	frids 0.5 hr/m^3, slab 0.4 hr/m^3	ditto	inc	−2700	ditto	inc		0
L03	Brickwork	Bricklayer	1100	9.00	9900	facings £250/th inc labourer							9900
L04		Br labourer	550	6.00	3300								3300
Totals			3150		23,100				−3000			−247	19,853

Figure 7.2 A specimen plant summary form.

PLANT SUMMARY

Project: Helix Laboratories, Westfield

Ref.	Resource type	Quant.	Unit	Output /hr	Output unit	Duration hrs	Rate £/hr	Totals in rates £	Sub-contract Name	Sub-contract Quote	Sub-contract Adjustment	Late quotations Name	Late quotations Quote	Late quotations Adjustment	Adjusted totals
PO1	JCB3, Founds	5800	m³	6	m³	967	14.00	13,533	} Thomas	28,500	-1833	F. Smith	26,350	-2150	26,350
PO2	Roller, founds	7200	m²	10	m²	720	5.00	3600	} ditto	inc		ditto	inc		
PO3	JCB3, hardcore	6000	m³	8	m³	750	14.00	10,500	} ditto	inc		ditto	inc		
PO4	Roller, hardcore	10,800	m²	20	m²	540	5.00	2700	} ditto	inc		ditto	inc		
PO5	20t tipper	5800	m³	8	m³	725	22.00	15,950	} Thomas	17,850	-4142	F. Smith	16,950	-900	16,950
PO6	Dozer on tip	5800	m³	24	m³	242	25.00	6042	} ditto	inc		ditto	inc		
Totals						3943		52,325			-5975			-3050	43,300

Completion of
estimate

Figure 7.3 A specimen materials summary form.

MATERIALS SUMMARY

Project:
Helix Laboratories, Westfield

Ref.	Resource type	Supplier	Quant. BOQ	Unit	Convent factor	Waste %	Total quant	Net rates £	Totals in rates £	Late quotations			Adjusted totals
										Name	Quote	Adjustment	
MO1	Hardcore	Target Stone	6000	m³	1.90	10.00	12,540	7.00	87,780				87,780
MO2	Type 1 subbase	Target Stone	7200	m³	2.10	10.00	16,632	6.75	112,266				112,266
MO3	Concrete 20N	MBC Concrete	250	m³		7.00	268	45.25	12,104				12,104
MO4	Concrete 30N	MBC Concrete	300	m³		5.00	315	52.60	16,569				16,569
MO5	Concrete sundries	Various	1	sum					4865				4865
MO6	Formwork	Various	1	sum					1850	F. Smith	1500	-350	1500
MO7	Facing bricks	Dexter Brick	55.4	th		6.00	59	335.00	19,673				19,673
MO8	Blocks 140 mm	Opus Supplies	6550	m²		7.50	7041	9.65	67,948		inc		67,948
MO9	Brick sundries	Various	1	sum					3860				3860
M10	Timber doors	Rosewood Joinery	65	nr		2.50			23,114				23,114
M11	Ironmongery	Lasermax	1	sum					11,480	Opus Supplies	11,020	-460	11,020
								Totals	361,509			-810	360,699

Figure 7.4 A specimen sub-contract summary.

SUB-CONTRACT SUMMARY

Project: Helix Laboratories, Westfield

Ref.	Trade	Quotations used in estimate		Discount	Net	Attendances			Inflation to form G	Late quotations			Adjusted totals
		Company	Quote			Lab	Pit	Mat		Name	Quote	Adjustment	
S01	Piling	BBS	40,079	2.5	39,077					Robinson	37,080	−1997	41,074
S02	Metalwork	No quote	69,610		69,610								69,610
S03	Structural steelwork	Steelcare	105,675	2.5	103,033					PCS	99,520	−3513	106,546
S04	Roof tiling	Richard Roofing	62,629	2.5	61,063				1500				61,063
S05	Leadwork	Richard Roofing	33,570	2.5	32,731				800				32,731
S06	Metal windows	Microtex	33,001	2.5	32,176								32,176
S07	Plastering	Addford	125,141		125,141								125,141
S08	Floor coverings	Floordex	29,844	2.5	29,098				700				29,098
S09	Painting	Clover	72,225	2.5	70,419				1700				70,419
S10	Plumbing	Sharpe & Baker	96,570	2.5	94,156								94,156
S11	Mechanical	Evanbridge	255,246		255,246	500	500	400	6600				255,246
S12	Electrical	Evanbridge	279,482		279,482	500	500	400	6700				279,482
S13	Lift installation	Expert Lifts	66,103	2.5	64,450	500		400		Clearframe	61,220	−3230	67,681
Totals			1,269,176		1,255,683	1500	1000	1200	18,000			−8740	1,264,424

91

3. The preparation of summaries of resources enables the estimator to list the most significant items, while management can assess the discounts, wastage factors and quantities allowed in the estimate. It is important that this reconciliation is made without relying entirely on the reports generated by a computer system.

4. Quotation totals from suppliers and sub-contractors can be checked against totals incorporated in the estimate.

Where computers are not used, unit rates are written into a copy of the bill of quantities and clerical assistants extend the rates and add page totals in order to arrive at a total for the project. In this case, verification of the calculations is carried out by a second 'checker' to ensure the total is correct.

For tenders based on drawings and specifications, an independent check on the principal quantities must be carried out in order to reduce the risk of a mistake, either leading to the winning of an undervalued bid or producing a high tender which may influence a client when giving opportunities to bid for future projects.

7.3   Price fluctuations

In times of relatively low and predictable inflation, contractors are expected to submit tender prices that remain fixed for the anticipated duration of the work. Fluctuations will be paid under a fixed priced contract in certain circumstances, for example, when materials are purchased during a period of compensable prolongation of the contract. For these purposes, the basic cost of materials are recorded in a list in an appendix to the bill of quantities.

Where the contract is to be fully fluctuating, then the method of establishing the increases or decreases in cost need to be checked. It is common for a formula method to be adopted and the rules for the application of the formula need to be studied carefully.

Where a firm price is to be submitted then an assessment must be made of the likely variations in cost during the proposed contract period. The tender works programme is an important tool in assessing the likely impact of price rises on certain elements of the project. There are some parts which will not need consideration such as provisional and PC sums, and firm price quotations which fully comply with the conditions of contract.

7.4   Risk analysis

Risks can be separated into known risks, known unknowns (uncertainties) and unknown unknowns (*force majeure*). The principal risks are identified which typically might fall into the following categories:

■ health and safety

■ environmental

■ design

■ construction

■ duration

■ innovation

■ existing conditions

- availability of resources

- external factors

- client

- consultants

- terms and conditions of contract.

These risks are then assessed for severity, i.e. impact if they were to occur, and probability of occurrence, the product of which gives a score which leads to the assignment of a high, average or low risk rating. Alternatively, they could be given a time and financial allocation so that the cost effect of a strategy is known. It is important to manage all risks having a high or average risk rating by putting into place an appropriate risk management strategy. Typical strategies include:

- risk avoidance such as a change in design which avoids a hazard

- risk transfer where a risk is passed on to another party such as a sub-contractor who will be better able to manage it

- risk mitigation where measures are taken to reduce the consequent impact of a risk to the project

- insurance against a risk (which is also a form of risk transfer)

- making cost and time contingency allowances to cover the impact of a risk.

The distinction needs to be made between transferring responsibility for the risk and the management of the risk. The contractor does not normally pass responsibility for the risk to a sub-contractor, only the management of the risk. Where the contractor retains ownership of the risk, they might consider also implementing a strategy to mitigate the risk, e.g. instigating changes to the scope or design of work, or making provision for re-sequencing of the tender works programme in order to be able to take the affected activity off the critical path, or to include some contingency period for its possible effect on progress.

These management strategies will be implemented by the tender team during the tender process so that the risks will have been to a large extent managed by the time the tender is to be finalised. This leaves any residual or newly identified risks to be managed as part of the tender settlement.

Risk assessment is best carried out as a group exercise or, in a risk workshop, involving the whole tender team. Different companies will have different attitudes to taking risk. Companies that are risk-averse may insist that all risks are reduced to a low level by tender submission; companies experienced at taking risks may tolerate some high risk items as long as time and money has been allocated in the tender to manage them, and as a reward for taking them.

Estimators generally make good risk managers as they develop a detailed understanding of the project as part of the estimating process, and have the necessary skills of analytical assessment and quantification to complete and implement the risk register at tender stage.

7.5 Alternative tenders

The estimator should note in which part the tender prepared may not meet the client's needs or where alternative approaches could give better value for money. It is acceptable to submit a compliant tender and to submit alternative tenders based on, for example, shorter construction periods or sectional completions for further consideration by the client.

7.6   Estimator's reports

It is essential that all pertinent facts which have an influence on the settlement are presented in a logical and structured sequence.

During the estimate preparation a large number of individual constituents will have been considered. These include:

- *The project and its construction characteristics*: identifying the market sectors, the size, nature and construction of the building and any special features.

- *Contract conditions*: identifying, among other things, the authorising body, e.g. Joint Contracts Tribunal, a statement on design responsibility and a period for which the contract is to exist, i.e. executed under hand or under seal.

- *Method statement*: describing the sequence, interface dates for commencement, completion and any phased handover or sectional completion and, any resource or preferential logic.

- *Tender works programme*: setting out the activities, durations and logic of the critical path network. Identifying the sequence, interface dates for commencement, completion and any phased handover or sectional completion dates.

- *Site factors*: identifying access and egress, ground conditions, environmental considerations and any constraints or controls to normal operational processes.

- *Sub-contractors and suppliers*: including details of any specified suppliers and products that may be from a single source.

- *Contractor's own preliminaries costs*: identifying plant, equipment and staffing numbers necessary to properly carry out and supervise the works.

- *Health, safety and environmental issues*: identifying those that may be relevant to the particular site, to particular components within the construction or constraints imposed by adjoining properties.

- *Cash flow*: which will indicate whether the project will be cash positive or whether financing will be a consideration.

- *Bonds and warranties*: identifying the level and period for bonding and the number and type of warranties to be given and by whom (usually including sub-contractors with a design input).

- *Insurances*: identifying the particular clauses of the contract which will apply to the works, together with a statement of requirements in terms if insurances are required to be held by the contractor.

- *Client issues*: which may identify their position in this particular marketplace together with trading history and payment performance.

- *Consultant issues*: looking at the size, capability and design experience of the particular practice in that market sector together with the numbers and experience of personnel the practice can dedicate to the project.

- *Quality*: this may be benchmarked in recognisable terms of specification or by example or sample of products and finishes.

Completion of estimate

- *Design*: looking at the complexity of the scheme, the design of interfaces and the available pool of skills currently available in the sub-contract market.

- *Pricing*: looking at the competition, the period for tendering and the existent situation in the marketplace in terms of available work and available resources to build.

- *Procurement matters*: relating to lead-in times for specific trades and/or goods where the design requires approval by the consultants before the purchase and fabrication of raw materials.

- *Risk/opportunity*: this is in effect a summary of the factors listed above which will be reflected in the settlement either in financial terms or by an initial qualification.

Some of these items will have been factored into the decision to tender and it is important to review these in that context in order to confirm that the final product still fits with the requirements of the company. For example, an organisation might have secured projects during the tender period that have an impact on the available resources making the current opportunity less attractive, i.e. the company may have to recruit staff for the project which would add a degree of uncertainty in outcome. All these items will be considered by management in relation to four key aspects of settlement.

Where there is a specific need for contractor design to be included on a traditional project, then an allowance for this risk in relation to the legal liabilities of the design needs to be included. A design portion supplement would be used in the contract to transfer design liability and this would give the value and scope of the work to be covered. The risks associated with the designed portion would need to be brought to the attention of the management for consideration at tender settlement stage.

Although the final decisions are made by management, those concerned with estimating, planning, construction and commercial management and purchasing must be encouraged to communicate the knowledge they have acquired throughout the estimating stage to the review panel. Their contribution may be by attending the meeting or reporting through the estimator.

8 Tender settlement

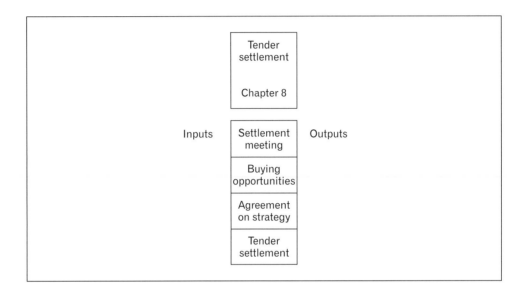

In large companies, the tender settlement stage often consists of two meetings: the first to review the estimate through a detailed examination of rates, quotations, methodology and sequence; the second would be for a director to receive an appropriate briefing and consider commercial matters before settling on the mark-up. The need for a formal and progressive approach in finalising or settling tenders should be regarded as fundamental to competent tendering.

An agenda for review meetings is indispensable. The agenda given in Fig. 8.1 can be used for most projects. For larger schemes, the agenda must be circulated to those required to attend.

An evaluation of alternatives, scope to improve profitability and risks which may be encountered should be considered at each stage of the estimating process and costs assessed in preparation for the settlement meeting. Clearly, in a strong competitive market, contractors need to find every opportunity to use products and processes which will enable the contract to support sufficient overheads and profits to maintain the company's objectives and satisfy the client by completing on time with the required specification.

Figure 8.1 An example of a standard review agenda.

	STANDARD REVIEW AGENDA	Project:	
The site	location		
	ground conditions		
	hazards and security		
	site visit report and photographs		
The parties	client		
	consultants		
The contract	form of contract		
	amendments		
	bonds		
	insurances		
	damages		
	payments and retention		
The estimate	labour		
	plant		
	materials		
	sub-contractors		
	PC and provisional sums		
The programme	method statements		
	specific requirements		
	tender programme		
Project overheads			
Final summary	estimate		
	firm price adjustment		
	review adjustments		
	value related items		
Mark-up	cash flow		
	scope		
	risk		
	head office overheads		
	profit		
	discounts (for main contractor)		
	VAT (where applicable)		

Tender settlement

8.1   Settlement meeting

When management considers risks at the settlement meeting an assessment is made of whether the priced documents comply with the tender requirements. The commercial and technical matters are checked to ensure, for example, that the construction method, sequence and timing and any assumptions are acceptable and it is assumed that the estimate has been correctly calculated with very few errors. In order to eliminate significant calculation errors an estimating department needs procedures for ensuring standards are maintained.

The mark-up shown on the final summary form (Fig. 8.2) comprises scope, risk, head office overheads and profit. Mark-up will be applied to all costs above this level. It can be assumed that provisional sums and daywork will produce their own contribution to overheads and profit. These latter items may be considered as part of the project opportunity.

At the settlement meeting there may be other decisions to be made before proceeding to the submission stage, including:

■   What documents will be submitted with the tender if nothing is specified in the invitation to tender? Some contractors may wish to prepare additional information such as a tender stage construction programme, company profile and printed brochures (if not previously provided during pre-qualification) whereas others reserve their efforts, and ideas, until they know that their tenders are under serious consideration.

■   If an alternative method (or design) is to be offered, then another price may need to be settled by the review panel.

■   Qualifications, which vary the requirements of the tender documents, are generally not permitted by clients. On the other hand, there may be circumstances where they cannot be avoided, and management must decide upon a suitable approach.

■   In order to provide a positive cash flow for the duration of the project it is important to decide how rates are to be apportioned in a bill of quantities to be submitted at the tender stage. Since any artificial alteration of prices will bring additional risks as well as opportunities, it is for management to agree a strategy.

■   Evaluation of the risks that have arisen during the estimating process. For example, has there been a change of personnel involved in the preparation of the estimate?

Where the bid is a two-stage process, i.e. where the selection of the contractor is made on limited criteria such as preliminaries and a percentage for overheads and profit, the final tender total may not be established until completion of the second-stage process. It is important that a further review be carried out at the end of this process to ensure that the targets and margins set at the initial tender meeting have been maintained through the detailed pricing stage and that the completed tender still meets management expectation. Depending on the state of advancement of the design and the situation with regard to planning approvals, the second stage may be for periods of up to a year during which time the needs of the company may well have changed. (See page 102 for an example of a final bid sign-off procedure.)

Tender settlement

Figure 8.2 An example final summary form.

FINAL SUMMARY

Project: Helix Laboratories, Westfield

	form	net costs lab	pit	mat	sub	site o'heads	PC & prov sums	sub-totals	totals
Own work									
Direct work	A–C	19853	43300	360699				423852	
Domestic subs & attendances	D	1500	1000	1200	1264424			1268124	1,691,976
Project overheads									
Time related	E	20000	51170	3175	100	116270		190715	
Fixed costs	E	11890	8140	2490	200	2840		25560	216,275
PC & provisional sums									
Nominated suppliers	F						18525	18525	
Nominated subs & attendances	F	300	300	600			139425	140625	
Statutory undertakers							0	0	159,150
Sub-total (estimate frozen)		53543	103910	368164	1264724	119110	157950	2067401	2,067,401
Adjustments									
Firm price	G			13200	18000	2000		33200	
Final review	H	–2300	3200	–28500	–22600	–11500		–61700	–28,500
Sub-total		51243	107110	352864	1260124	109610	157950	2038901	2,038,901
Value related items									
All-risks insurance, Empl & Pub Liab						11940		11940	
Water for the works						see proj o/h		0	
Performance bond						3450		3450	15,390
Professional fees									
Total net costs		51243	107110	352864	1260124	125000	157950	2054291	2,054,291
Percentage breakdown	%	2	5	17	61	6	8	100	
Mark-up							%	£	
Scope								–25000	
Risk							3	60879	
Head office overheads							5	104508	
Profit							5	7019	
Sub-total								147,407	2,201,698
Provisional sums from form J									96,000
Dayworks from form J									60,800
TENDER TOTAL excl VAT								£	2,358,498

Signatures:

8.2 Buying opportunities

With many contractors boasting high levels of work won under preferred status and not in competition, the same is often true for key sub-contractors and suppliers. They must demonstrate value for money but main contractors may not be in a position to take buying opportunities.

8.3 Agreement on strategy

There are two parts to the strategy: tender preparation strategy and selling/profit recovery strategy. Companies may have a particular preference; however, one approach is for the strategy to be agreed at the tender kick-off meeting. Each member of the bid team should write their respective strategy: estimator–pricing strategy; architect–design strategy (if appropriate); commercial manager–commercial and risk strategies; supply chain manager; planning engineer and so on.

8.4 Tender settlement

When contractors set tender margins, they consider all the points discussed in the final tender review meeting (see Fig. 8.1). In addition, they may be influenced by the competition and their past tendering history and by budgets set by clients or clients' consultants, if known. It is clear that neither of these criteria will help achieve optimum margins in a highly competitive market. It is difficult, however, to avoid low margins where competition is strong but, if prices fall below economic levels, it is time to look for other market sectors where the competition is less fierce.

The settlement meeting should review all aspects of the bid preparation and be controlled by an agenda. The settlement will focus on the confirmation of the true net estimate by adjudication of the individual elements prior to setting the tender margin:

■ overhead recovery

■ profit in accordance with stakeholder expectation

■ review of risks

■ opportunities arising in carrying out the contract, e.g. improvement in programme performance, in construction method, in cash management, in the buying and in the prospect of additional work.

This settlement of the estimate and, its conversion to a tender, is the responsibility of management. The accountability of the estimator should be limited to the proper preparation of the predictable cost of a project. It must not be considered that the estimator's responsibility is to secure work for the company.

Appendix

Sample Form

COMPANY NAME
TENDER SETTLEMENT

MINUTES OF SETTLEMENT MEETING

PROJECT:

DATE AND TIME OF MEETING:

LOCATION OF MEETING:

PRESENT:

APOLOGIES FOR ABSENCE:

DISTRIBUTION:

Agenda:

1. Project details

2. Construction methodology

3. Construction sequence and timing

4. Preliminaries also pre-construction costs

5. Sub-contractors also consultants

6. Own work

7. Contract conditions

8. Financial statement on client

9. Risk analysis also opportunities analysis

10. Tender clarifications

11. Settlement details including bonds and insurances

12. Tender submission

SAMPLE TEMPLATE FOR MINUTES OF SETTLEMENT MEETING AT TENDER SETTLEMENT STAGE

Action

1.00   PROJECT DETAILS:

2.00   CONSTRUCTION METHODOLOGY:

3.00   TENDER WORKS PROGRAMME:

4.00   PRELIMINARIES:

5.00   SUB-CONTRACTORS' WORK PACKAGES:

6.00   OWN WORK PACKAGES:

7.00   CONTRACT CONDITIONS:

8.00   FINANCIAL STATEMENT ON CLIENT:

9.00   RISK ANALYSIS:

10.00   TENDER CLARIFICATIONS:

11.00   FINANCIAL DETAILS:

11.10   Profit:

11.20   Contractor Contingencies:

11.30   Client Contingencies:

11.40   Daywork and Provisional Sums:

11.50   Fixed Price Allowance:

12.00   TENDER SUBMISSION REQUIREMENTS:

Tender
settlement

SAMPLE FORM

PRE-SUBMISSION SIGN OFF

Tender No:	Project:		Contract No:
Scope		**Approval/Comment**	
Agreement to scope of work			
Technical review of drawings and information			
Reconciliation of price to scope of work			
Design review			
Identification of bid qualifications			
Contract terms and conditions			
Warranties, PCG and bonds			
Payment guarantee			
Status of major work elements			
Staff availability			
Mobilisation status			
Any other business			
Agreed by:		Date:	

Tender settlement

9

Post-tender activities

Post-
tender
activities

Chapter 9

Inputs

Queries
and
adjustment
of errors

Outputs

Tender
results

Negotiations

Letters of
intent

Contract
award

Handover
to delivery
team

Project
control

Feedback

Performance
data

9.1   Queries and adjustment of errors

The estimator should be prepared to respond to any request for further information or a notification that the submitted tender contains errors in computation. For errors, two alternative courses of action are possible:

■   The tenderer will be given details of such errors and afforded the opportunity of confirming or withdrawing the offer. The estimator will need to refer to management when, the extent of the computation errors has been determined, for a decision as to whether to confirm the original tender figure or withdraw the tender. Under English Law it is possible to withdraw the tender at any time before its acceptance.

■   The second option is that the tenderer is given an opportunity of confirming their offer or of amending it to correct genuine errors. If the contractor elects to amend their offer and the revised tender is no longer the lowest, whatever then becomes the lowest tender will be examined in more detail and may be preferred.

The estimator must consult with management to establish whether to amend the tender figure or to confirm the original offer, once the extent of the computation error has been determined.

In both situations, the estimator or signatory to the original tender must be prepared to endorse the appropriate tender documents to note the acceptance or change to the tender. Such amendments must also be endorsed by the employer in the event that a contract is subsequently awarded.

There are many ways to correct errors in a bill of quantities. The most common is to recalculate the bill and make an adjustment on the final summary page, either by changing the total for preliminaries or introducing an adjustment item. Since the latter amount must be applied to all valuations, it is preferable to ask the contractor to change one or more items in the preliminaries in order to bring the total to the tender figure.

9.2   Tender results

Standard practice recommends that tenders should be opened promptly, and all but the three lowest tenderers should be informed immediately that their tenders have been unsuccessful. The second and third lowest tenderers should be informed that their tenders were not the most favourable and that they might be approached again if it is decided to give further consideration to their offers. They must be notified at once when a decision to accept a tender has been taken. Other forms of procurement would have specific procedures for tender closures.

Once the contract has been let, every tenderer should be promptly supplied with a list of the tender prices. The estimator (or chief estimator) should record the results in the tender register (Fig. 9.1) and, if required, in a tender performance report, illustrated in Fig. 9.2.

When the result of a tender is known, the tender performance form (Fig. 9.2) should be completed, where possible with a comparison made with the lowest (or the accepted) tender. Tender results can be reported at regular management meetings so that all those who contributed to the estimate can assess their performance. Suppliers and sub-contractors who submitted quotations should be notified of the results as promptly as possible.

Tender documents should be carefully filed for possible future reference; except drawings and bills of quantities which are of little use once the contract has been awarded to another contractor and should be returned or destroyed. Quotations received from suppliers and sub-contractors can provide a useful guide to prices for other tenders, but great care is needed to ensure that the specification and the nature of the work are the same.

Estimators must monitor their performance with an analysis of results recorded over a period of consistent estimating. If data are available from a large number of tenders, it is possible to evaluate tender performance in relation to: types of work (new build or refurbishment), clients (public, private), procurement routes (traditional, design and build, trade packages), and value of projects. With this information, decisions can be made about where to concentrate estimating resources for further tendering.

Post-tender activities

Figure 9.1 A specimen tender registration form.

	TENDER REGISTER		Start _____ Finish _____			

Tender number	Client	Title	Tender		Result		Notes
			Date	Sum	Yes % low	No % high	
10166							
10167							
10168							
10169							
10170							
10171							
10172							
10173							
10174							
10175							
10176							
10177							
10178							
10179							
10180							
10181							
10182							
10183							
10184							
10185							
10186							
10187							
10188							
10189							
10190							
10191							
10192							
10193							
10194							
10195							
10196							
10197							
10198							
10199							
10200							

Figure 9.2 A specimen tender performance report.

	TENDER PERFORMANCE REPORT	

No	Client	Title	Our tender	Lowest tender	% over lowest	% over mean

Post-tender activities

Client consultants are prepared, in most circumstances, to offer either formal or informal feedback to an unsuccessful tenderer. This is an excellent practice as it assists the bidder in the improvement of his tendering to achieve a better understanding of how bids are received and reviewed.

9.3   Negotiations

When the lowest (or the best value) tender received exceeds the client's budget, changes should be negotiated with the most favoured tenderer. This process is either through recommendations from the contractor for value engineering cost savings, or design changes which reduce the scope or specification of the works. If significant changes are proposed to a scheme, two or (at most) three tenderers may be asked to re-tender in competition thereby retaining the lowest market prices for the amended project.

If there has been any delay in acceptance, enquiries should be made to ensure that there have been no changes which might necessitate reconsideration of the tender.

9.4   Letters of intent

Once an acceptable offer has been made, the awarding of a contract should be in clear terms, stating that the offer has been accepted. There should be no reason to use a letter of intent, which can lead to confusion about how payments will be made for abortive work.

However, it should be noted that, where the pre-order of materials is essential, this may be initiated by the client or under a limited letter of intent covering sub-contract design and/or procurement.

Whichever method is used, the estimator must obtain approval from management that the wording can be accepted in order to start the planning, as well as the mobilisation of resources. A standard award form (Fig. 9.3) should be used to provide an overview for other departments and, a facility for the responsible director, to indicate his acceptance of the terms and provide the authority for the construction team to incur expenses.

9.5   Contract award

When a tender has been successful, it is necessary to take certain action before the actual contract documents can be signed. This action will be carried out by various people within the company's organisation but it is recommended that it should be co-ordinated by the estimator who was responsible for the original submission. The documents must be checked thoroughly to ensure they reflect the exact content of the documents used to prepare the tender.

Contract documents should be checked to establish that:

- the drawings are those which were circulated with the tender documents; if they have been revised to produce a construction issue, they cannot be used as contract documents

- the dates, penalties and particulars given in the appendix to the conditions of contract are those stated in the tender document

- the submission bills have been copied correctly.

Figure 9.3 A specimen contract award form.

	CONTRACT AWARD	Tender number
		Contract number

Project particulars			
Project title		Client	
		Project manager	
Project address		Architect	
		Quantity surveyor	
Project description		Structural engineer	
		Services engineer	
Award letter	Date	Planning supervisor	
Acceptance			
Letter of intent		Principal contractor	
Other			
Contract			
Start		Contract value	
Completion date		Bond value	
Duration		Release date	
Early action			
External	Date		*Internal*	Date	
Acknowledgement letter			Mobilisation programme		
Pre-start meeting			Health and safety plan		
			Handover meeting		
			Start-up meeting		

Distribution

☐ Commercial   ☐ Operations director

☐ Purchasing   ☐ Area manager

☐ Insurance   ☐ Contracts manager

☐ Accounts   ☐ Business development

☐ Safety   ☐ Marketing

☐ QA   ☐ PR

Signed   Date

(Director)

A standard form may be used to confirm that these checks have taken place. In larger organisations, this form will accompany the documents until they are signed by the company's directors and returned to the client's representative.

9.6 Handover to delivery team

The following information, as appropriate, must be made available to those who will be responsible for construction and purchasing:

- correspondence
- form of tender
- priced submission bill
- bill of internal allowances
- build-up of rates fully adjusted for review changes
- tender works programme and method statements
- quotations
- analysis of quotations
- project overheads
- estimator's notes
- site visit report
- further information received after tender submission (if any)
- tender documents.

It is recommended that an internal pre-contract handover meeting be held with all persons concerned in attendance, so that details are fully discussed and the distribution of information can take place at the meeting. The handover meeting is an opportunity for detailed reporting and a discussion on decisions made at the time of estimating concerning methods of construction, site layout, organisation, suppliers and sub-contractors. These decisions should never be made binding on those to be responsible for construction. It is important that the reasons underlying such decisions and choices should be explained fully. However, decisions made at the time of tender cannot be ignored, unless it can be demonstrated that a better method of operation is possible.

9.7 Project control

Cost coding systems

It is recommended that a fully operational cost control system be introduced on all successful contracts. Although it is outside the scope of this Code, it should be noted that the collection and collation of data is a specialised function involving the use of statistical techniques. It is important that during construction information should be obtained on the labour and plant cost of major items or stages of work, on the quantities of materials used, and the cost of attendance for sub-contractors. This information is essential for control purposes and guidance in future estimating. These records should be accurate and give all details including those of the work, the position on the site of the operation, gang sizes, levels of incentives and bonuses being paid, plant, access and weather conditions. This information is needed so that it may be interpreted correctly for future use.

Estimators should ask for this information so that they can adjust their databases if necessary.

Where a substantial proportion of work is assigned to sub-contractors, a table should be drawn up showing the estimator's allowances and actual prices agreed with them at the time of placing their contracts. In particular, the estimator needs to know the accuracy of risk, buying gains and firm price provisions in the tender. Where the company operates a cost:value reconciliation model in their commercial management process, there should be a link to those in procurement identifying site performance.

Equal importance should be attached to the reconciliation of the expenditure of preliminaries sums, particularly where these may not be under a sub-contract form, and consequently not subject to any specific or detailed audit trail.

Record keeping

Effective records of resources used and, productivity achieved are the skeleton on which effective project planning is built for the tendered works, including delay and disruption claims that support any changed works. Unless good records are kept of what was carried out, where, by whom, when, and how that differed from that which was tendered for, no claim for additional costs or time can be proved.

Records kept on paper are of little use in managing progress. They cannot be searched, retrieved, compared or accumulated. Accordingly, contractors should consider either having paper records re-keyed at weekly intervals into a database, or have the data entered directly into a database that will facilitate the preparation of fast, accurate, progress reporting.

Project scheduling

A critical path network is an essential prerequisite of effective time management. The master programme should be constructed as a fully logically linked critical path network, that will react dynamically to change. The purpose of the exercise should not be the drawing of a pictorial diary, so much as a tool to predict consequences.

The master programme should be supplemented by short-term, fully resourced logically linked programmes, integrated into the master programme to avoid foreseeable adverse consequences.

Updating and revising the master programme

Some contracts require the master programme to be updated whenever there is any departure from the approved programme. Others require the effect of an extension of time to be identified and some require nothing at all. Irrespective of what is required by contract, it is in the contractor's interests to regularly update, review and revise the master programme with the best information currently available at the time.

It is also vital that any departure from the tendered work be contemporaneously identified on an update programme, so as to calculate the time and hence time-related costs of the departure, and to provide the necessary information for an extension of time and/or compensation.

Whenever the master programme is updated or revised, a copy of it should be sent to the contract administrator for his information and/or acceptance, depending on the contract form.

It is never advisable, nor is it good practice, to work to a works programme different from the published master programme; such practices only lead to poor time management and difficulties in substantiating claims.

9.8 Feedback

Those who were responsible for preparing the estimate should make periodic visits to the site during the progress of construction. This will enable them to objectively assess the soundness of decisions made during the estimating process or subsequent site decisions where changes were considered necessary to improve performance and/ or outputs. In addition, they should maintain a constant record of the conditions which prevail on sites.

9.9 Performance data

At the completion of a project, the estimate and tender should be reconciled with the final cost and final account and the reasons for the results assessed in detail for guidance in future estimating.

It is also beneficial if the estimator is sent a brief report indicating the performance of all sub-contractors and a recommendation on whether they should be employed for future projects. Larger companies will operate a database of resources which will indicate a review system whereby the site teams will report on sub-contractor performance under a number of predetermined headings (see Fig 9.3). This, coupled with performance monitoring at tender stage, will provide worthwhile feedback and guidance for future bidding. With the current culture of a sub-contractor operation, the maintenance of performance monitoring of the supply chain is becoming a key aspect of the procurement activity.

Post-tender
activities

10 Alternative estimating techniques

This Code has dealt in depth with the production of an estimate and tender where a detailed design and a bill of quantities is available, and the constituents of rates (labour, plant, materials and sub-contracts) are priced individually and summarised separately for the review meeting and handover.

With the many different procurement approaches now being used there is often the need for the contractor to be involved at the early design stage and for their input into the approximate estimating process.

At an early stage in a project's development, approximate estimating techniques are employed in order to set budgets and assess the feasibility of a scheme and, where necessary, gain funding. During design development, a cost plan is an important way of producing a design within the client's budget.

10.1   Approximate estimating

Approximate estimating techniques depend on historical cost data being available from previous similar schemes. In practice this information is analysed and applied in three ways:

- functional unit or unit of accommodation method
- superficial (floor area) method
- elemental cost plan.

Functional unit or unit of accommodation method

This method uses the unit cost from earlier schemes. The unit cost is calculated by dividing the total cost of building by the number of functional units. For example, the functional unit for a motorway hotel might be a single room, the cost of which includes a small proportion of the front office and corridor space. A hospital bed may be the basic unit used in a general hospital and a car space in a multi-storey car park. Clearly, this approach can be used to set broad yardsticks in setting budgets for construction, but it must be recognised that there is a significant difference between this budget and the final cost of construction. This is because the nature of the site varies, the client's brief has to be developed, incoming services are not always available close to the building, and the shape and specification for the building affects the price, as will the commercial aspects of the contract when the project is tendered.

Figure 10.1 An example elemental cost plan.

	ELEMENTAL COST PLAN			Project	Helix laboratories	
				Type	Two-storey offices	

		KB Electronics Gross floor area (m²) 2,900 Elemental costs			Helix Laboratories Gross floor area (m²) 3,850 Elemental costs		
ref	Elements	£/m²	totals	notes	£/m²	totals	notes
1	Substructure	91.94	266,625	RC pads	128	492,800	piled founds
2A	Frame	90.19	261,563		90	346,500	
2B	Upper floors	48.88	141,750		49	188,650	
2C	Roof	68.08	197,438		68	261,800	
2D	Stairs	19.20	55,688		19	73,150	
2E	External walls	84.38	244,688		84	323,400	
2F	Windows and external doors	62.84	182,250		63	242,550	
2G	Internal walls and partitions	41.90	121,250		42	161,700	
2H	Internal doors	8.90	25,819		9	34,650	
3A	Wall finishes	18.04	52,313		18	69,300	
3B	Floor finishes	82.63	239,625		83	319,550	
3C	Ceiling finishes	23.86	69,188		24	92,400	
4	Fittings and furniture	12.80	37,125		13	50,050	
5A	Sanitary appliances	7.56	21,938		8	30,800	
5C	Disposal installations	5.94	17,213		6	23,100	
5D	Water installations	8.15	23,625		8	30,800	
5E	Heat source and space heating	70.41	204,188		70	269,500	
5G	Ventilation and cooling	27.93	81,000		28	107,800	
5H	Electrical installation	139.66	405,000		140	539,000	
5J	Lift installation	–	0		31	119,350	two lifts
5K	Security alarms	36.83	106,819		37	142,450	
5K	Fire alarms	14.49	42,019		14	53,900	
5N	Builder's work in connection	9.48	27,506		9	34,650	
	Net building cost	974.09	2,824,875		1,041.00	4,007,850	
6A	Site works	142.27	412,594		142	546,700	
6B	Drainage	32.64	94,669		33	127,050	
6C	External services	9.66	28,013		10	38,500	
	Net trade total	1,158.67	3,360,150		1,226.00	4,720,100	
7	Preliminaries	89.80	260,415		90	346,500	
7	Design fees	60.52	175,500		61	234,850	
7	Statutory fees	14.90	43,200		15	57,750	
7	Pre-start costs	–	–		0	–	
7	Inflation	–	–		120	462,000	
7	Overheads and profit	74.48	216,000		74	284,900	
7	Contingencies	46.55	135,000		47	180,950	
	Budget total	1,444.92	4,190,265		1,633.00	6,287,050	

Superficial (floor area) method

This method uses historical data from earlier comparable schemes in terms of the cost per square metre of floor space. This is a popular method readily understood by developers and the building team, as relatively few rules are needed to apply this technique. The floor area of a building is defined as that measured at each floor level using the internal dimensions, without making deductions for internal walls or stairs. As with any approximate estimating technique, a number of adjustments are needed to take account of location, specification, degree of complexity, size, shape, ground conditions and number of storeys.

In order to assess these factors, reliable historical costs are needed from a variety of buildings within each building category. A separate assessment must be made of external works, main services and drainage which can all vary substantially depending on the nature and location of the site.

Elemental cost plan

An elemental cost plan can be produced from a preliminary building design. Again the method depends on reliable data being available from comparable projects where the actual costs for each building element is known. Figure 10.1 shows how prices from a bill of quantities for a similar office building have been set against a standard list of elements. In this example, the floor area of the previous building was 2900 m^2. It is shown that the breakdown of costs for a 3850 m^2 building can be calculated simply by applying the earlier proportional cost in each element to the second building. A number of adjustments are easily made, such as the introduction of a lift and a piled foundation.

10.2 Composite unit rate estimating

A large part of an estimator's time is devoted to calculating unit rates for items in a bill of quantities. In addition to analytical unit rate pricing, which is dealt with in Chapter 5, the estimator will use the following techniques;

- spot items
- operational estimating.

Spot items

These are operations which are difficult to break down into discrete items of work in a bill of quantities. For example, the demolition of small buildings or the formation of openings through walls is priced by looking at the extent of the work during a site visit.

For estimating purposes, spot items may be treated in several ways:

- Approximate quantities can be taken off and unit rates used to calculate a lump sum estimate for the item.

- The description within the bills can be analysed into its constituent operations and trades and an estimate of the cost made for each.

- When the description within the bill is analysed into constituent operations and found to have a predominant trade, then a gang or operational assessment can be made on a time, plant and material basis so that the overall cost can be calculated.

In all three methods the cost of labour, plant and materials should be separated in accordance with the general principle described for unit rates in Chapter 6.

The work must be inspected thoroughly at the site visit and, where necessary, construction method must be established and documented. Adequate allowances must be made for storage, temporary work, including supports, access, double handling, small deliveries, making good and reinstatement. Transport can be included in the item, but will usually be included in project overheads.

If bill descriptions are not clear, or if further information or measurements are required, it may be necessary to revisit the site.

Operational estimating

This system is adopted when the estimator needs to consider the overall duration of an operation and its interrelationships with other trades. This is the case with civil engineering construction or the earthworks and concrete elements of a building project. In these cases, it is unrealistic to look at a single unit of work and wrong to assume the total cost of an operation is the product of the unit rate and quantity.

For example, a contractor may make an assessment for laying precast concrete manhole rings on the basis of the number a drainage gang can fix in one day. For a 2.10 m diameter manhole and 600 mm high units, the estimator might assume that 15 units can be handled, lifted into position and securely bedded in one day. If a project has 25 precast concrete units, an allowance of two days may be needed because it might be difficult to deploy the plant elsewhere for a small part of the second day. It is unlikely that a simple unit rating exercise would have included an allowance in this way for standing time.

Operational estimating depends on a careful study of how a section of work will be carried out in practice. It is difficult, for example, to price the fixing of roof trusses without looking at working methods.

Case study 1

A new dental surgery has a rectangular plan shape, 35 m long, with 55nr timber roof trusses above first floor level spanning 8.50 m between wallplate supports. The estimator has drawn up a list of resources for fixing trusses as follows:

Mobile crane (1nr)	1 day		@	380.00	=	380.00
Banksman (1nr)	8 hrs		@	10.00	=	80.00
Carpenters (2nr)	16 hrs		@	15.00	=	240.00
	for 55 trusses					£700.00
	for 1 truss					£12.73

In this case a method statement was not produced. But for more complex construction operations, for example, more planning is needed together with method statements.

Case study 2

A large distribution warehouse has a floor slab with 850 m^3 of concrete to be cast in continuous pours with all joints and reinforcement to be introduced during the casting operation. For the method chosen 300 m^3 of concrete will be placed each day.

After a detailed review of the design and discussions with the consultants, it was agreed to replace steel reinforcement with fibre-reinforced concrete. The estimator and planning engineer drew up a list of resources and durations, as follows:

Trade foreman	(1nr)	3 days x 9 hrs @ £14.00	378.00
Gangers	(2nr)	3 days x 9 hrs @ £12.00	648.00
Operatives	(10nr)	3 days x 9 hrs @ £11.00	2970.00
Carpenters	(2nr)	3 days x 9 hrs @ £15.00	810.00
Laser screeder	(2nr)	3 days x 9 hrs @ £14.00	756.00
Driver	(1nr)	3 days x 9 hrs @ £15.00	405.00
Dumper	(2nr)	3 days x 9 hrs @ £5.00	270.00
Drivers	(2nr)	3 days x 9 hrs @ £14.00	756.00
Power floats	(2nr)	3 days x 9 hrs @ £5.00	270.00
		Total cost for 850 m^3	£7263.00

In order to insert a unit rate in the bill of quantities, the total cost for placing concrete is divided by the quantity.

Unit rate for labour and plant = £7263/850 m^3 = £8.54/m^3

It is not difficult to use computer-aided estimating software for operational estimating. A list of resources can be produced for an item of work and applied to the total quantity; the computer automatically divides the total cost by the quantity in order to arrive at a unit rate. The last example would be input in the following way:

Concrete slab:	850 m^3
Pricing quantity:	850 m^3
Total cost =	£7263.00
Unit rate =	£8.54

Code	Description	Unit	Quantity	Rate
LFT	Trade foreman	hr	27.00	18.00
LGA	Ganger	hr	54.00	12.00

etc.

10.3   **Approximate/builder's quantities**

Builder's or approximate quantities are needed where the other approximate estimating techniques do not produce sufficient information for a reliable budget. The term 'approximate quantities' is used for a number of arrangements, the most common being:

- A shorter bill of quantities with composite items. An item for external walls, for example, would include both skins of masonry, forming the cavity, wall ties, plastering and pointing. In this case an approximate bill of quantities is produced and priced with rates taken from a number of sources including previous bills of quantities, price books or guide prices from specialist trade contractors and suppliers. The accuracy of this method will depend on the extent to which the design has been developed. If approximate bills of quantities are to be used for cost planning during the design stage, they should follow an elemental bill format giving estimated costs for each building element.

■ A contractor's bill of quantities produced from drawings and specifications that includes fewer ancillary items than are required by the standard method of measurement. The rules for measurement, such as deducting openings in walls, are often ignored, as it is assumed that the over-measure will account for the extra labour and increased waste on materials. Computer-aided estimating systems provide a quick method for creating a bill of quantities. The estimator selects items from a library of descriptions which were previously priced. Resource costs can be changed when material quotations are received.

■ In order to start a contract early, a bill of quantities (often from another project) can be used to establish a tender price. The JCT Standard Building Contract with approximate quantities is the variant for this arrangement.

10.4 Cost planning

Increasingly, contractors (and consortia in public sector projects) are being asked to develop designs and methods which meet affordability targets. Figure 10.2 shows stages in the pre-contract phase of a construction project. The cost plan is developed as a vehicle for continuous reporting and monitoring costs as the design advances.

Figure 10.2 The development of cost studies at pre-contract work stages.

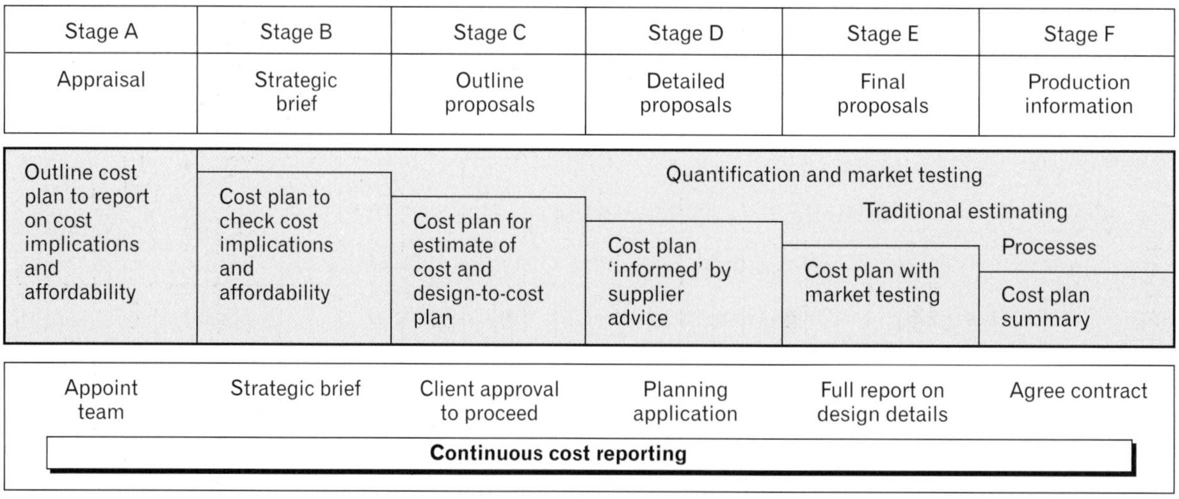

A simple 'outline' cost plan can be produced at the earliest appraisal stage and this will be expanded into a 'first-pass' cost plan at Stage B 'Strategic brief'. An example of a simple cost plan is shown in Fig. 10.3. The standard building costs have been taken from average rates; the on-costs are usually known, which means that the main effort is put into the abnormals – those items which are site specific or part of an enhanced design.

Figure 10.3 An example first-pass cost plan for student accommodation.

Capital Cost
First-Pass Cost Plan (m² method)

Proposed student accommodation at Southampton

Overview of capital costs	schedule areas	%	£	£/m2
New build	17,509 m²		12,568,310	717.82
Building totals	17,509 m²		12,568,310	717.82
Abnormals				
Enabling works – fencing			450,000	25.70
Utilities upgrade			425,000	24.27
Site works and drainage			245,055	14.00
Drainage			95,600	5.46
Buildings and abnormals	net cost		13,783,965	787.25
Preliminaries total	15,599,314	13.17	1,815,348	103.68
Construction risk on all works		2.92	402,492	22.99
Design fees/charges exc M&E		8.37	1,153,718	65.89
Pre-start costs		5.08	700,225	39.99
			17,855,749	
Inflation on net build cost + prelims		10.00	1,559,931	89.09
			19,415,680	
Margin on gross turnover		7.00	1,461,395	83.47
			20,877,075	
106/278 Infrastructure contribution	net/gross	1.51		
VAT @ 17.5%			20,877,075	
Construction out turn total		£	**20,877,075**	**1,192,36**
To start 2Q10				

Student accommodation *Benchmark data from Exeter*

Overview of capital costs	drawn areas	%	£	£/m2
New build	18,470		13,258,135	717.82
Refurb games room	235		207,505	883.00
Building totals	18,705		13,465,640	719.90
Abnormals				
Enabling works – demolition			151,698	8.11
Extl engineering services			218,849	11.70
Incoming gas main			43,209	2.31
Site works and drainage			132,806	7.10
Drainage			137,295	7.34
Buildings and abnormals	net cost		14,149,495	756.46
		13.17	1,863,489	99.63
		2.92	413,628	22.11
		8.37	1,183,910	63.29
		5.08	718,182	38.40
			18,328,704	
		incl	18,328,704	–
		7.51	1,377,297	73.63
			19,706,001	
		1.39	–	–
VAT @ 17.5%			19,706,001	
Construction out turn		£	**19,706,001**	**1,053.52**
To start 2Q09				

Figure 10.4 An example top-down target cost plan.

	Top-down ... Design-to-cost plan	Southampton student accom block

Number of bedrooms		710	
Gross internal floor area		17,509	
Capital cost affordability			**20,000,000**
On-costs — Margin		−0.07	−1,400,000
Sub-total			18,600,000
Inflation on build cost and prelims		−0.1000	−1,494,396
Pre-start costs		−0.0508	−670,808
Design fees and charges		−0.0837	−1,105,249
Contingencies/risk		−0.0292	−385,583
Preliminaries		−0.1317	−1,739,083
Net build and abnormals cost			**13,204,881**
Abnormals — Enabling works – demolition and fencing			−450,000
Utilities upgrade			−425,000
Site works			−245,055
Drainage			−95,600
Net buildings cost			**11,989,226**

	m²	£/m²	
Std costs — New build	17,509	**684.75**	11,989,226
Refurbishment	–	–	–
Target net cost			**11,989,226**

Once the budget has been established with the client, a top-down design-to-cost plan can be produced (see Fig. 10.4) so that the design team understand the cost limit. On-costs are generally fixed by the terms of the framework agreement and abnormals need to reflect site conditions. The main building costs are then presented in a design-to-cost report which can be in the form of an elemental cost plan or as a list of principal quantities and descriptions. Figure 10.5 is a typical design-to-cost report aimed at designers who need costs translated into simple quantities and statements.

Case study

A developer builds and operates accommodation for students in many towns and cities in the UK. In order to ensure a strong business case, a unit cost per study bedroom was established. The value of a student bedroom often depends on local demand, land values and finance costs.

This case study is for a 710-bedroom development in Southampton. The business plan has been optimised to produce a reasonable return on the investment and indicates an affordability construction cost of £28,170 per bedroom. The total cost limit has therefore been set at £20,000,000.

Figure 10.5 An example elemental design-to-cost cost plan.

		Elemental design-to-cost cost plan				Project	Southampton
						Type	Student accom

		Southampton Design-to-Cost Cost Plan			Notes	
		GIFA	**17,509**	m²	Net internal area = 14,704m²	
		elemental costs			Height 6 storeys	
ref	Elements	totals	£/m²	%		
1	Substructure	913,795	52.19	5%	Pad foundations and ground fl slab 200 mm thick	
2A	Frame	1,136,684	64.92	6%		
2B	Upper floors	716,468	40.92	4%	Precast concrete upper floors (excl screeds)	
2C	Roof	451,732	25.80	2%		
2D	Stairs	80,541	4.60	0%	Finishes in element 3B	
2E	External walls	1,329,809	75.95	7%	95% facing brickwork and 5% curtain walling	
2F	Windows and external doors	257,732	14.72	1%	Punch windows 1340 mm square	
2G	Internal walls and partitions	1,007,643	57.55	5%		
2H	Internal doors	584,100	33.36	3%		
3A	Wall finishes	359,285	20.52	2%		
3B	Floor finishes	451,032	25.76	2%		
3C	Ceiling finishes	293,451	16.76	1%		
4A	Fittings and furniture	1,069,275	61.07	5%		
5A	Sanitary appliances	949,863	54.25	5%		
5B	Services equipment	–	–	0%		
5C	Disposal installations	75,989	4.34	0%		
5D	Water installations	189,622	10.83	1%		
5E	Heat source and space heating	377,494	21.56	2%	Convector heaters to rooms and corridors	
5G	Ventilation and cooling	91,397	5.22	0%		
5H	Electrical installation	607,037	34.67	3%		
5J	Lift installation	379,945	21.70	2%	Four passenger lifts	
5K	Security alarm system	224,465	12.82	1%		
5K	Fire alarm system	207,131	11.83	1%		
5N	Builder's work in connection	234,796	13.41	1%		
	Net building cost	**11,989,288**	**684.75**	◄—	**Target** 11,989,226 684.747	
6A	Site works	245,000	13.99	1%	Soft landscaping £85,000	
6B	Drainage	95,600	5.46	0%		
6C	External services	425,000	24.27	2%		
6D	Enabling works – demolition + fencing	450,000	25.70	2%	Fencing 355 m with two single gates	
	Net trade total	**13,204,888**	**754.18**			
7	Preliminaries	1,739,083	99.33	9%		
7	Design fees	1,105,249	63.12	6%		
7	Statutory fees	–	–	0%		
7	Pre-start costs	670,808	38.31	3%		
7	Inflation	1,494,390	85.35	7%		
7	Overheads and profit	1,400,000	79.96	7%		
7	Contingencies	385,583	22.02	2%		
	Budget total	**20,000,000**	**1,142.27**	**100%**		

A construction company, who worked closely with the developer, was asked to produce an initial cost plan for the scheme to test the viability of the project and later to produce cost studies to enable the design team to design within the budget. The contractor's strategy was as follows:

1. Convert the number of bedrooms into a net building area (the sum of all the rooms). Add space for corridors and common areas to arrive at a gross internal floor area (GIFA).

2. Visit the site to gain an understanding of site constraints and ground conditions. These abnormal works items often have a significant bearing on the practicality of the development. On the other hand, some abnormal site costs can be used to negotiate a better (cheaper) land deal.

3. Obtain standard building costs for the superstructure works. These square metre rates are drawn from an analysis of a similar project in a database of building costs. Contractors are in a unique position to gather current trade package costs from projects on site. Package costs are then converted to element costs in the form of a standard elemental cost plan.

4. Agree on-cost items with the client, such as margin and design fees. In this example, these costs were agreed earlier as part of a framework agreement.

5. Produce an outline cost plan, Stage A, with: standard costs, abnormals and on-costs. An example is given in Fig. 10.3.

6. If the total construction cost is affordable, then an elemental cost plan will be prepared using data from a similar scheme.

7. Prepare a design-to-cost report, comprising the elemental cost plan and clear statements of what the elemental costs represent in terms of principal quantities and components.

8. Attend design team meetings, monitor the emerging design and suggest ways to remain within the overall target cost.

9. Obtain advice on market prices from specialist sub-contractors.

10. As the scheme design reaches detailed proposals, produce a bill of quantities from scaled drawings and carry out further market testing to confirm the assumptions made in the cost plan.

Using this process, the contractor can provide continuous cost reporting, from the scheme appraisal stage through to agreeing a contract sum.

10.5   PFI and whole-life costing

Contractors who are involved in PFI projects are now having to produce whole-life cost plans as part of their tender bid to support their planned maintenance requirements for the 20 or 25 years of the contract.

With the need for accurate historical cost data on maintenance, organisations such as the Building Cost Information Service (BCIS) and the Building Research Establishment (BRE) have started to collect a wide range of data. The collection of data involves costs in itself and until there was a market for the data very few independent organisations were prepared to invest in it.

It is not the function of this Code to give detailed guidance on how to prepare a whole-life cost plan but the bid management team should be aware of the requirement for this as part of the tender submission.

Alternative estimating techniques

In the early 1990s the British government formulated and awarded specific Design, Build, Finance & Operate (DBFO) projects to modernise the UK's ageing road infrastructure; these in the main were the forerunners to 'fully' privately financed projects (PFI).

Also there were early Build, Own & Operate (BOO) projects and Build, Own, Operate & Transfer (BOOT) projects but the DBFO projects were more numerous.

It allowed various road infrastructure projects to be built and operated by private companies but more importantly, financed by private capital with the borrowings remaining 'off balance sheet' for the government.

However, these private loans were normally 100% underwritten by the government and repaid over a concession period of (initially) 20–25 years through 'shadow tolls' for traffic usage: in effect the government was buying upgraded assets on hire purchase.

These projects tended to be multimillion pound projects such as major new motorways and strategic bridge links such as the Second Severn Crossing and the Dartford M25 Toll Bridge.

John Major's government then launched the Private Finance Initiative (PFI) in the mid-1990s. PFI contracts were of similar format but were used to include renewal and upgrade of other key government-funded facilities.

There were two main forms of contract which passed down different levels of risk to the private sector, namely:

■   'availability' such as schools, prisons, hospitals

■   'full risk' such as water projects, light rail and private roads.

The 'availability' model followed very much the same format as DBFO in that 'availability for use' payments were made, much in line with 'shadow tolls', during the concession period.

Where the 'full risk' was passed down, the lengthy concession period allowed revenue to be generated by the private companies directly from the service provided and was used to finance and service their equity and loans.

These 'full risk' projects were not underwritten by the government.

Structure of PFI/PPP company

In order to allow investment in PFI/PP projects and to allow the effective risk transfer to safeguard the private investors, a new format of legal company had to be established; Special Service Companies (SPC) or Special Service Vehicles (SPV) were established as the main contracting company.

These SPC/SPVs became not only responsible for the construction of the specific project but had to operate the facility for up to 30 years before handing back to the government.

A typical structure of a SPC/SPV is shown in Fig. 10.6. Within this structure, there needed to be formal contracts to ensure all parties understood their legal responsibilities and then were held to account to deliver to the specific requirements and performance criteria.

Figure 10.6 A typical structure of a SPC/SPV.

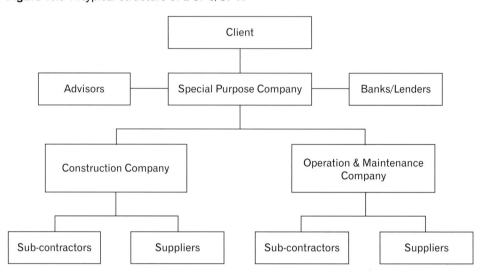

As in the main the companies forming the SPC/SPV were the same as being involved in the Construction and O&M Companies, there needed to be formal 'arms-length' stand-alone contracts between the different legal entities.

Estimating for a PFI project

There is no doubt a considerable amount of work for bidders who participate in the PFI route. The primary areas of consideration would include:

■ understanding the requirements and expectations for a 'cradle to grave' performance

■ preparing and or advising on the schedule of activities

■ preparing early cost plans and cost forecasts (from the feasibility stage)

■ developing target cost plans against the scope and specification and advising on 'affordability'

■ developing and advising on value-based solutions for optimum whole-life cycle

■ developing and advising on the financial model for the whole-life cycle.

Similar to projects in public sector, the process of tendering for PFI is dictated by the EU Procurement Directives and would relate to the amount of information available and the design stage reached.

Box 10.1 Example of a pricing strategy for a PFI hospital scheme

---

CB Construction Limited

Pricing Strategy for Stansford NHS Trust PFI Hospital

This is a combined PITN/FITN scheme leading to selection of preferred bidder. At the start of the 26 weeks tender period no design has been produced and the Trust's Public Sector Scheme will be issued shortly.

There are three stages as follows:

1. Target setting

2. Cost control and net construction cost

3. Submission documents

1. TARGET SETTING

Affordability

Assessment by Consortium following discussions with Trust finance director.

E.g: £180,000,000

Gross Internal Floor Area

Set area target.

E.g: £180,000,000 divide by £2750/m^2 = 65,500m^2

Area breakdown

Break down area into departmental gross, communications and plant.

E.g: 14% on dept gross for communications; 11% for plant

So departmental gross	=	52,400m^2
Communications	=	7,340m^2
Plant	=	5,760m^2

Schedule of accommodation

■ Health planner produce departmental schedule of accommodation. E.g. 52,400m^2.

■ List departments in same order as Trust.

■ Estimator check maths in schedule of accommodation.

■ Produce strategy for closing gap between affordability and drawn area.

New build and reconfiguration

Assess where accommodation can be provided in retained estate.

Potentially convert Pathology into University labs and Fred Jones ward into outpatients.

Produce categories of reconfiguration: heavy/medium/light or more refined.

Tactics might be:

■ 90% of cost for heavy refurb/reconfiguration.

■ 55% of cost for medium refurb/reconfiguration.

■ 20% of net cost for light refurb.

continued

Box 10.1 Example of a pricing strategy for a PFI hospital scheme

---

First-pass cost plan for run of financial model

Produce a single sheet cost plan:

- Buildings priced on £/m^2 costs
- Add abnormals
- Add equipment
- Add infrastructure charges
- Add typical mark-ups for risk, fees, inflation and margin.

Target cost plan for cost control

- Elemental breakdowns for target affordability
- Research elemental cost plans for typical hospital buildings
- Check elemental costs against similar projects

2.   *COST CONTROL AND NET CONSTRUCTION COST*

Cost control

- Design to cost document – issue target cost plan and elemental costs to design management.
- Attend design meeting and advise on compliance with target costs.
- Do sufficient taking off to check elemental costs.

Procurement

- Market testing to extent possible within design programme constraints.
- Consider: site works and frame; fitting and furniture; equipment.
- Advise cost team on current budget costs.

Programme for capital building price

	Activity	from week	to week
1	Issue of ITN documents	1	1
2	Project appreciation	1	2
3	Affordability target	1	3
4	Gross internal floor area (GIFA)	3	3
5	Area targets for communications and plant	3	3
6	New build/re-configure splits	3	5
7	First-pass cost plan for Financial Model	5	5
8	Target cost plan	2	6
9	Attend design meetings and monitor costs	4	24
10	Report on cost plan at regular intervals	4	24
11	Draft cashflows for interim runs of financial model	4	24
12	Take off and price external works and shell	16	20
13	Advice from trade for major packages	16	22
14	Measure final GIFA on submission drawings	22	24
15	Complete estimate cost plan	20	24
16	Complete site overheads book (project overheads)	23	24
17	Settlement meeting	24	24
18	Final cashflow to include in financial model	25	25
19	Insert costs in submission documents	25	26
20	Submit bid	26	26

continued

Alternative estimating techniques

Box 10.1 Example of a pricing strategy for a PFI hospital scheme

---

Net construction cost

- Produce data sheets from previous hospital tenders/contracts.

- Compare Trust brief with developing scope to identify over-provision

- Identify scope items that exceed the benchmark costs.

- Modify elemental costs where necessary.

- Use costs current at time of tender. Inflation can be dealt with separately.

- Produce project-specific project overheads.

- Produce project-specific equipment schedule from room data sheets.

- Use spreadsheet format that accords with submission requirements.

3. ***SUBMISSION DOCUMENTS***

Drawings and specification

- Check submission drawings meet requirements of cost plan.

- Check submission specification meets requirements of cost plan.

Costs workbook

1. Single spreadsheet workbook with elemental cost plans, and summaries of risk, fees, prelims, cash flow and inflation.

2. Schedule of equipment costs.

Produce early cost plan, by week three if possible.

Your financial advisor will need some costs in order to set up a spreadsheet model.

Pick an architect who fully understands the market sector.

For example, up-to-date experience in education is vital for a school project. There is a temptation to choose an architect or engineer because he is well known by the client, or is willing to work at risk in the early stages.

Decide who is responsible for each aspect of cost.

Show on a chart who is pricing: capital maintenance, routine maintenance, decanting, life-cycle fund, new furniture, up-grading existing building stock, etc.

Do people understand their roles?

Completing the price

Design fees

Infrastructure charges

Risk

Inflation

Margin

continued

Box 10.1 Example of a pricing strategy for a PFI hospital scheme

---

Cashflow forecast

Input to financial model, may also be required for submission.

Life-cycle costs analysis (capital replacement costs)

What information is needed for the life cycle cost model?

Tender submission

Including FM, site developments, variant bids and finance.

---

11 Contractual arrangements

This Code of Practice has focused so far on a conventional lump sum (CLS) form of procurement, using a contract such as JCT Standard Building Contract. There are other procurement approaches that are widely used and therefore need to be considered. Some of the forms of contract in current use would include:

■ design and build (D&B)

■ management contracting

■ construction management

■ two-stage tenders

■ turnkey projects

■ PFI contracts

■ cost-plus contracts

■ term contracts

■ serial contracts

■ framework agreements

■ partnering agreements

■ sub-contracts

■ trade contracts.

While it is not the aim of this Code to cover all forms of contract and procurement, nevertheless this chapter outlines three typical alternative forms of procurement (in addition to CLS which demonstrates how the bid management process is adapted to meet the different requirements): design and build contracts, two-stage tenders, as well as PFI contracts.

11.1   Design and build contracts

Figure 11.1 A bid management flowchart for a D&B tender.

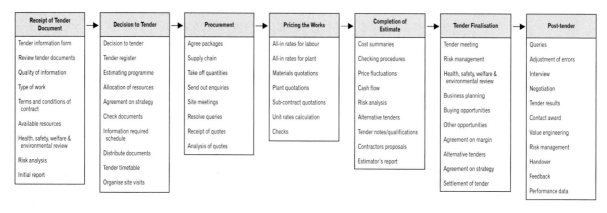
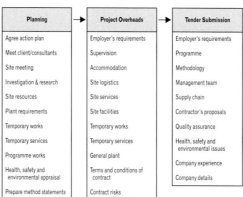
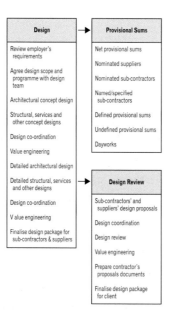

The D&B form of procurement is very popular with some clients and used extensively outside the UK. The benefits are usually perceived to be:

■ more economical projects, as the contractor can apply his expertise to use the most cost effective design solutions, materials and methods to achieve the client's requirements

■ the client passes some risk to the contractor

- the design period and lead time can be shortened by overlapping design with construction where the tender process permits it.

These features of D&B bring significant challenges to the pre-construction process. The drive to find economical design solutions can involve the contractor in taking risks by using cheap materials or plant which subsequently may prove to be unreliable. Design risk is passed on to the contractor which for some projects can prove to be very considerable. Compressed design and construction programmes can lead to significant management problems with design delays tending to be primarily at the contractor's risk.

Contractors must bring additional resources and procedures into the pre-construction process to meet these challenges. Risk management is of particular importance. For major D&B projects it is wise to employ an independent risk manager with a background in design risks to join the team. Design can be carried out by an internal design department, external design practices or by specialist sub-contractors. The best solution will depend on the company and the project. Only major companies with a regular flow of D&B work can afford to maintain a design department so sub-contracting of design and drawing production is a very common solution.

Employer's requirements (ERs)

The scope of documents included contractually under ERs can be quite varied and include meeting obligations under agreements to lease and the like.

Contractor's proposals (CPs)

CPs are the contractor's response to these and generally are the vehicle for proposing changes to the issued enquiry.

Consultants and the design and construct process

The design team for a D&B project will usually be assembled specifically to work during the tender period on a selected project. The make-up of the team will vary depending on the nature of the project, but a typical one would include:

- the contractor, who is represented by estimator, planning engineer, project manager designate (to cover the buildability practicalities), design and technical manager (to co-ordinate the design)

- architect

- civil and structural engineer

- mechanical and electrical consultant or, alternatively, a mechanical and an electrical sub-contractor (the latter would provide services design either using their in-house resources, or by employing an external mechanical and electrical consultant themselves)

- others (if relevant).

It is normal for one of the contractor's representatives to lead the team, often referred to as the 'bid manager', and for the architect to assume the role of 'lead consultant'.

Selection of consultants   Assuming that the contractor is at liberty to choose his own consultants (this is not always the case as they can be novated, and this is dealt with separately below), then the following selection criteria will be considered:

- fee basis during the tender stage (no win – no fee)
- resources readily available
- good track record on similar projects
- competitive fee for the post-tender stage
- be pro-active, and contribute to the team
- cost conscious and able to produce cost-effective solutions
- individual advantages (e.g. know the building if the project is a refurbishment, have a good relationship with the client, etc.).

Not a contributing factor to a particular project but, often a consideration imposed by a contractor's business development plan, is the consultant's ability to introduce the contractor to other prospective business opportunities. There may also be contractual requirements that require the consultant to have certain qualifications, belong to specific professional institutes, carry certain professional indemnity insurances or enter into collateral warrantee agreements and, obviously, these also need to be confirmed with the consultant.

Novated consultants

To provide continuity throughout the inception, design and construction process, it is common for a client to employ consultants to develop a scheme from inception through to tender stage, paying their fees during this period, and then to 'novate' them to the successful design and build contractor. The contractor then employs them for the remainder of the design and the construction period.

Contractors should be guarded in their dealings with these consultants during the tender period, as the consultant's first loyalty will be to the client, and also the consultant would be in direct dialogue with the contractor's competitors.

D&B: design team meetings

The estimator's role in these should be pro-active. The estimator should challenge the designers to produce cost-effective solutions, and not simply to take on a clerical role and price which is presented to them. Where the estimator sees excessive cost, the team should be made aware of it, as it is generally the most cost-effective scheme that wins, *not* the most extravagant. The estimator should also make the team aware of the importance of maintaining the key dates on the tender preparation programme for the provision of design information. This ensures that sub-contractor and supplier enquiries can be issued on time, and that bill of quantities preparation can be managed within agreed periods.

Value engineering should be an area where *all* team members are encouraged to participate. Each idea should be given a unique identification number and recorded on a list. The estimator can then approximately cost each, discarding those which are not effective, and proceed to produce detailed costs for the remainder. The estimator can then present the list at the tender settlement, and decisions can then be made regarding inclusion or otherwise in the eventual tender.

Due to the wide range of specialists attending and, the invariable complexities of the project, design team meetings need to be structured, and conducted to a timetable.

Defining the consultant's role and agreeing fees

Standard fee bid forms

The relevant professional institutes already have standard forms for fee bidding which act as tick sheets to clearly define roles and responsibilities. Consultants will be well versed in agreeing briefs and fees based on these and it is recommended, therefore, that they are used as the basis for the relevant engagements.

Copyright issues precludes the provision of an example in this publication; however, copies should all be readily available via the appropriate consultant.

Defining the details and responsibilities
The prospective consultant must be given a clear and precise brief regarding what services are to be provided and, what responsibilities are to be carried. Fringe matters, such as the number of copies of each drawing to be issued, the consultant's involvement with on-going site inspections, frequency of site meetings and who is to attend, will all affect the fee and should be detailed in the brief.

Dates for issue of design information, which will dictate the levels of resources that the consultant will need to consider, together with details of warrantee requirements, and levels of professional indemnity insurance, also need to be defined. In effect the consultant should be clear as to their role and associated responsibilities.

If the consultant is to provide services during the tender period on a 'no job – no fee' basis, this should also be clearly stated, and the value and recovery of those costs, should the tender be successful, should be incorporated into the post-tender fee structure.

It is desirable to have a 'design responsibility matrix' listing all elements that have design input, with a clear statement of who has what level of responsibility for them under one of the following categories:

- total responsibility

- primary responsibility, with input from those with contributory responsibility

- contributory responsibility, led by those with primary responsibility.

Sharing bills of quantities

Most D&B tenders have strict and well-defined employer's requirements which cover most of the project. They usually contain drawings and specifications describing all aspects of the superstructure, elevations, layouts, materials to be used, finishes, services, fittings, etc., leaving only structural elements, substructure and drainage to be finalised by the tendering contractors. The client does not seek a design for the building as such, but knows the requirements precisely. They will probably have employed an architect to prepare drawings and a specification, a services consultant to do likewise for the mechanical and electrical services, and a quantity surveyor to prepare tender documentation and solicit tenders.

The client seeks simply to place the design and quantification responsibility (together with the attached and associated risks) on to the successful contractor.

In this situation, it is quite common for several tendering contractors to jointly share a quantity surveyor to prepare bills of quantities for the standard elements, thus allowing them each to concentrate their own resources on the non-standard elements. Provided that the process is managed with integrity, there are numerous advantages:

- The cost of bills of quantities preparation for the unsuccessful contractors is limited to what they spend on billing the elements which they design.

- There is less risk regarding quantification, i.e. missed or wrongly measured items, as the bills are prepared by a specialist who is skilled in the process and has procedures and checks. As this risk is often priced or considered when assessing margin, this reduces the tender price.

■ Sub-contractors and suppliers receive similar bills of quantities extracts to price from the various tenderers and thus price the job once rather than be faced with pricing it several times. This alleviates confusion as often, where there are several different sets of bills of quantities extracts have been produced, sub-contractors and suppliers tend to price the first one that they receive and copy that by return to anyone who subsequently sends them an enquiry.

■ Queries to the client during the initial phase of tendering can be raised once from one source. This can be done by the shared quantity surveyor via one of the tenderers if protocol is to be preserved, or by the shared quantity surveyor directly if the client is willing. The latter obviously saves time and is more accurate as it is a direct enquiry. It also allows the quantity surveyor to query detail and intent, which may have been overlooked during the preparation of the tender documents, thus producing common and equitable solutions.

The shared quantity surveyor produces a bill of quantities, but has *no involvement or influence* on how it is priced. The fee for this, provided that it is a fair reflection of the cost of providing the service, does not amount to an attempt to fix a tender price and, therefore, cannot be construed as collusion.

Payment

The normal procedure for payment is that the successful contractor pays the fee once the contract is awarded. It should be written into a 'winner pays' agreement that in the event of the scheme not proceeding to contract with one of the participating contractors, then the fee will be shared equally between all the participants.

In order to clearly define the agreement, method of payment, details of the service and timescale, etc., there should be a formal, and detailed, proposal from the quantity surveyor to the tendering contractors wishing to participate, which should be formally accepted by each of them.

Quantity surveyor's brief

To start the process, one contractor must act as agent for all by allowing the quantity surveyor access to the tender documents to assess the extent of the billing required. In addition, all participants must agree to a clearly defined 'quantity surveyor brief' for the bills of quantities preparation and date(s) when the completed document will be delivered.

Quantity surveyor selection

There are no rules applicable to this. However, it is obviously important that the selected quantity surveyor is able to deliver the goods. Other issues to be considered include previous track record related to the type of work to be billed, knowledge of the site, previous experience of being a shared quantity surveyor, ability to call in help if the workload suddenly increases (addendum letters), and a proven commitment to achieve target dates.

■ The shared quantity surveyor must be agreed by all the participating contractors.

■ If several are short listed, then a fee bid should be organised, allowing all sight of the documentation available and giving a clear brief.

■ It must be emphasised that the fee tendered at this stage must account for an element of the unknown, regarding the issue of addenda by the client or his professionals, as no increases will be permitted once the fee has been agreed.

One contractor should take the lead in briefing the common quantity surveyor on behalf of all participants and issuing documents and addenda. However, all may have access with regard to queries, etc.

Design by specialist sub-contractors can prove to be a successful solution as not only is the design carried out very economically, but also the design risk is passed on to the sub-contractor. Where the client issues a highly detailed employer's requirements brief, incorporating a full concept design, the sub-contractor design solution can work well. There is a residual design risk with the contractor in this circumstance which may still make it advisable to retain competent design consultants to review sub-contractor's designs for compliance with the employer's requirements and for co-ordination of design with other sub-contractors and designers. The contractor's design team for a D&B project will typically include design consultants, specialist sub-contractors and possibly suppliers. This team will need managing, as will the design process and programme.

Design management, like project planning, scheduling and risk management, is an expertise in itself. For simple projects or, design development for specialist packages for conventional lump sum tenders, e.g. contractor's design portion supplement under the JCT Standard Building Contract, contractors may believe it possible for estimators and construction managers to manage design. Where the element of design becomes significant, even only for co-ordination of design of specialist sub-contractors, design managers or technical services managers must be employed by the contractor to manage the process and the consequent risks. A major D&B bid management project team is likely to include:

- bid manager or project manager in the lead role

- estimator

- planning engineer

- risk manager

- design manager

- architect

- structural engineer

- technical services manager

- mechanical engineer

- electrical engineer

- CDM co-ordinator or safety advisor.

The pre-construction process for a D&B tender looks very similar to the process for a conventional lump sum tender, with the addition of design and design review stages. The practice is however quite different in being more complex, taking longer and using a far greater level of resources. There are complex inter-dependencies, which we have not attempted to show, between for instance design and procurement, which require time and management. D&B tenders are inherently more time consuming and costly to prepare and have far higher risks than conventional tenders. Consequently, the estimator must make specific cost provisions, not only for design and design management, but also for all risks that are consequential to taking responsibility for design.

There are many hybrids of D&B, some of which include:

Contractual
arrangements

- Contractor's Designed Portion Supplement (CDPS) – this procedure is commonly used under the JCT Standard Building Contract to apportion design responsibility through the main contractor to specialist sub-contractors. This commonly applies to services installations but can also be applied to joinery, cladding and many other elements of the contract. The extent of CDPS design will indicate the requirement for in-house design and design management required. These resources must be costed and the risks must be identified, managed and priced.

- Detail and build – not a defined contractual arrangement, more a situation that arises from the way in which a D&B or conventional contract is amended or applied. A client has employed a design team to develop their requirements to a detailed concept stage and then wishes to engage a contractor to complete the detail of that design and build the project. The advantages to the client are getting a best-value solution to the detailed design and construction, passing on a large amount of risk to the contractor and maintaining maximum control over the design. The contractor must make a careful assessment of the residual design required, how best to complete this design and what are the risks and how to manage them. As with CDP, the design responsibilities and risks can by-and-large be passed on to specialist sub-contractors; however, there will be significant design management and coordination responsibilities and there can lead to large residual risks to manage and cost.

- Novation of designers – is in effect a derivative of detail and build. A client engages a design team to prepare concept designs as part of his employer's requirements for a D&B tender, making it a condition of the tender that the contractor then engages the same designers to complete the detailed design and often also to supervise the works. He gets the same benefit of passing on design risk to the contractor while maintaining a very large degree of design control. This contractual arrangement can lead to a conflict of interest for the designers and additional risks for the contractor. The design contracts originally entered into by the employer are novated to the contractor on his appointment via novation agreements, which are usually drawn up and agreed at the time of the original design appointments. The stated intention is for the designers to fulfil their design responsibilities to the contractor as if the original design agreements had been between designer and contractor. This is best reinforced with the addition of a design warranty from the designers to the contractor. A conflict of interests can arise where the client expects the designer's post-novation to represent his interests in, for instance, maintaining design and construction standards. This might be resolved by defining what those standards are in the employer's requirements, thereby making them a requirement under the contract in any case.

- Turnkey projects – this takes D&B a stage further from the client's point of view, giving him a facility to which he just has to add his management and staff to make it operational. This could be for instance a hotel, so that the contract would entail in addition to D&B construction the provision of furniture, equipment, plates, cutlery and possibly staff training. Once again the contractor's involvement is increased with a consequent increase in the time and resources required and the risks to manage and cost. Projects of this nature are commonly a joint venture between a construction contractor and a facility operator who can bring the additional knowledge and expertise required. Joint ventures entail establishing from the outset a joint management team with a clear structure and decision-making powers to prepare and submit the tender, then to undertake the contract. Responsibilities, costs, risks and rewards must be carefully apportioned from the outset.

- Two-stage D&B (2ST D&B) – review 2ST contracts in detail below, however there are some particular aspects of 2ST D&B worth mentioning. First, there are significant benefits to the client in that the design process may be compressed even further by integration into the contractor's pre-construction activities, saving time, and design risks will be passed on to the contractor at the second stage. There is a major benefit to the contractor in 2ST D&B, in that the design, costs and programme are not firmed up until the second stage, giving the contractor the maximum opportunity to manage and minimise his risks during pre-construction. 2ST D&B is particularly applicable to fast-track projects and projects where the client's requirements must be developed along with the design. There is a major inherent problem with 2ST D&B. D&B relies on there being a stated client's requirement in order to establish a fixed point from which to start a contract. In 2ST D&B, this may not be possible in the first stage, or the requirements may be so vague as to be difficult to apply. This issue must be addressed by careful drafting of the tender documents to enable effective change control at the first and second stage.

D&B is an increasingly popular form of procurement to clients who see it as a route to get a good value project, delivered as quickly as possible. Many contractors have successfully carried out many profitable D&B contracts, however, many others have made some very costly mistakes either at tender stage or on-site. The management challenges, particularly during the pre-construction stage, are considerably greater than for a conventional contract and place a considerable responsibility on the estimator to prepare a sound and professional estimate, and to risk manage the tender process.

11.2   Two-stage tenders

A two-stage tender (2ST) process can be applied to many different forms of contract including CLS and D&B. The intention is to appoint a contractor at an early stage in the project to join the project team to carry out

- cost planning
- project planning and scheduling
- risk management
- design development
- buildability
- value management
- procurement
- site investigations and surveys
- enabling works.

The aim is to improve the overall outcome of the project in terms of improved cost and value, reduced overall project timescale, reduced risk, in-depth appreciation and achievement of the client's requirements and better quality in terms of function and finish. It is particularly appropriate for high risk, fast-track and innovative projects where the contractor can play a major role in managing out risk during the pre-construction process.

Figure 11.2 A pre-construction flowchart for a two-stage tender.

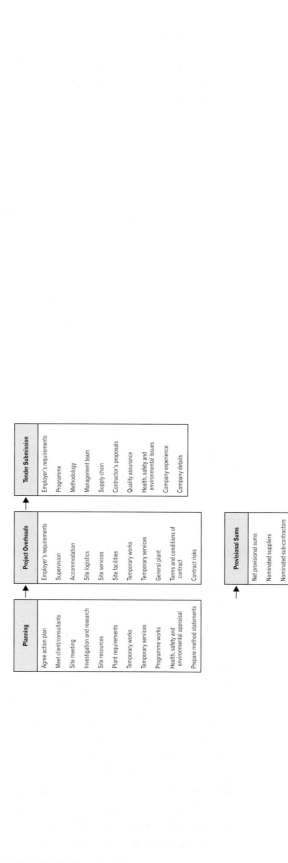

The first-stage tender in its simplest form involves the selection of a contractor on the basis of tenders for a combination of a pre-construction fee, contract preliminaries and percentages for overheads and profits (O&P) on construction costs. In addition, a schedule of rates may be priced by the tenderers to be applied to all or parts of the works. Cost comparisons are achieved by applying the tenders to a cost model for the contract with either an estimated prime cost to apply O&P percentages to or a schedule of provisional quantities to apply to the schedule of rates. Cost is seldom the only or indeed the most important criteria when selecting a contractor for the second stage of the contract. This form of procurement relies on using the contractor's expertise and resources to the best advantage during the pre-construction and construction phases in order to obtain best value for the client.

Therefore, selection criteria can be ranked thus:

1. Expertise and experience of the contractor's proposed team.

2. Methodology for managing and building the project.

3. Project planning and scheduling proposals.

4. Expertise in risk, design and supply chain management.

5. Contractor's track record on similar projects.

6. Contractor's financial background.

7. Approach to health, safety and environmental issues.

8. Procedures for quality assurance.

9. First-stage tender sum.

The tender submission document covering the above issues is therefore a critical part of the first-stage tender and is likely to be followed by post-tender interviews of the contractor's team. Therefore at this stage, estimating becomes a small part of the process. The estimator's expertise will be needed for the second stage to procure packages and prepare the second-stage tender. So his curriculum vitae will form part of the submission, as may a cost plan or a review of the client's cost plan.

The 2ST incorporates stages familiar from CLS and D&B single-stage tenders as the end result will be a lump-sum fixed price and time contract. The difference is in the open-book nature of the process as the procurement and costing are done in conjunction with the client's consultants. Tender packages, a design release, procurement schedule and the list of tendering sub-contractors, are discussed and agreed in the first instance. The contractor and the client's quantity surveyor (PQS) draw up and assemble the tender packages including sub-contractor preliminaries, design information, schedules of work or bills of quantities and tender conditions and instructions. The contractor will organise site meetings and mid-bid meetings with all the sub-contractors. Sealed tenders are often returned to the PQS. The contractor will then hold post-tender interviews with favoured sub-contractors and final selection is made by the whole team on the basis of a report and recommendation by the contractor for each package. Where schedules of rates are used, the works are measured and priced by the contractor for review and agreement by the PQS. Rates may apply to the whole contract, however, they are more normally only applied to trade packages to be carried out by the contractor's own labour which may included general building, carpentry and decorations. In this way costs and timescales are agreed on a package by package basis and the 2ST sum and tender works programme are agreed.

The second-stage process may be carried out and completed before the works are started on site so that the client has complete cost and time certainty before committing himself to the contractor for anything more than his pre-construction fee. For fast-track projects such as office fit-out contracts this will not be the case as one of the main advantages that the client is seeking is to take possession of his new offices as soon as possible. It is quite common for such contracts to start on site within weeks of the first-stage appointment on works covered by rates in the first-stage tender, with subsequent packages being agreed on a just-in-time basis to get to site at the earliest possible moment. Procurement and appointment is done on rolling short-term programmes driven by the release of design information and the construction programme for the required completion of each sub-contract package on site within the parameters of the master programme. This process requires very careful project planning and management, good working relationships between the contractor, PQS, planning engineer and project scheduler and designers, and professional non-bureaucratic methods. Cost certainty for the client may not be achieved until the works are well advanced on site which means that he is carrying design, cost and time risks for a large part of the project. However, there may be no alternative if he wants to achieve an early occupation date. In the City of London, office rental costs are high, and although they are often mitigated by rent-free periods to enable tenant fit-out projects to be completed, the times allowed are short and seldom give much time, if any, for design and lead in, so fast-track 2ST contracts are very attractive financially despite the consequent risks.

There is a close resemblance between 2ST, management contracting and construction management contracts. The two main differences are:

■ Contractual arrangements with the client and supply chain. Under construction management contracts the client contracts with specialist and trade contractors and the construction management contractor is employed in a consultancy role. Under management contracting contracts the contractor contracts with trade contractors on the client's behalf. Under 2ST the contract becomes a CLS or D&B contract after the second stage, with works being carried out by sub-contractors and operatives employed by the 2ST contractor in the normal way.

■ The risk allocation between client and contractor. Under construction management contracts, the client retains virtually all the usual risks of the employer in regard to each package contractor. Under management contracting contracts, the contractor takes on more of what are usually the employer's risks as he employs the trade contractors; however, there is no fixed cost or time agreement for the contractor to be held to and he is usually paid a percentage of the value of the works packages. Under 2ST contracts the end result is a lump sum fixed price and time contract with the contractor eventually taking on board most project risks from the client.

There are also derivatives, including guaranteed maximum price and incentive clauses, used to modify management contracting and construction management contracts. These seek to transfer more risks to the contractor and reduce the control that the client may otherwise exercise.

11.3 PFI contracts

PFI contracts take the progression from CLS to D&B to turnkey projects to their logical conclusion. This is often called a cradle to the grave service. Particular projects will differ; however, the overall arrangement can cover procurement of land, finances, design, construction, operation, facilities management and ultimate disposal. Such projects are vast in their scope and tremendously complex in their execution. In Fig. 11.3 we show a highly simplified view of the process in order to locate the pre-construction process and the role of the estimator in the overall project. This has also been introduced in Chapter 10.

The PFI pre-construction process is inevitably long and costly with large multi-profession teams including architects and engineers, accountants and solicitors along with construction professionals. The process needs project management, cost management, design management and planning over a long period, which leads construction companies to make long-term secondments of key personnel to PFI project teams. Estimators can take key roles in the process from outset to completion.

Having received an initial enquiry from the client, a feasibility study must be carried out by the contractor or tendering organisation which will include a risk assessment and an assessment of the cost of the project. On receipt of the formal tender documents the project risks will be reassessed and the project cost plan and the tender cost budget will be developed in detail in order to give realistic cost targets for the major elements of the project and to agree fixed tendering costs for all tender cost heads. Cost and time planning are essential from the outset to make financial control possible as the project develops. Having made the decision to tender, familiar organisational tasks will be required to ensure good communication and document control. This will also be the stage at which the tender team, including specialists such as accountants, solicitors and property consultants, will be confirmed and strategic decisions will be made. A tender preparation programme and project procedures will be agreed with all the parties to enable efficient management of the tender. The next phase of the tender is the development of detailed proposals for all elements of the project from land purchase to operation of the facility. This phase can be particularly complex with many inter-related and dependent activities which can be challenging, particularly where teams have not worked together or not worked on PFI projects before.

The all-up tender will be based on a cost model for the facility covering costs for:

■ setting up the PFI operating company

■ land or premises purchase

■ capital funding

■ design

■ construction

■ operation

■ facilities management and maintenance

■ disposal or termination.

The estimating team will be required to prepare costings in the traditional fashion for the construction of the project, taking into account all design elements and design risks. In addition, there is likely to be the need to prepare whole-life costings including maintenance costs. As with D&B tenders, all the elements of the tender proposals and costings must be co-ordinated and brought together in a single tender proposal document. In any complex project the risk and consequences of making errors are high so robust procedures for supervision and checking of every stage of the process are crucial. A thorough risk analysis must be

Figure 11.3 A pre-construction flowchart for a PFI contract.

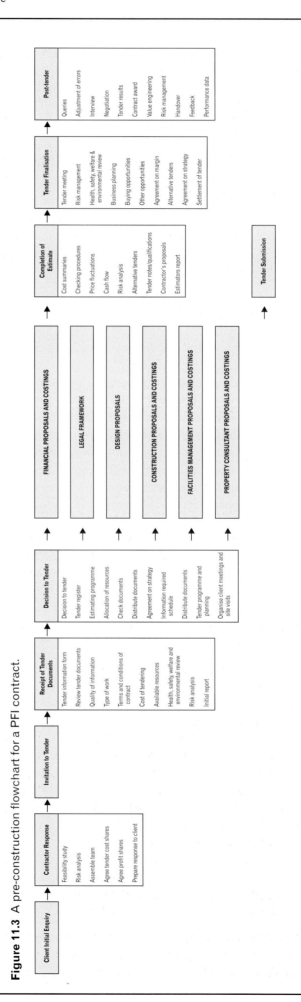

carried out for every project element as the tender develops to enable risk management measures to be adopted.

The cost model must then be converted to a tender by adding an allowance for overall or residual risks plus overheads and profits as agreed by all parties with a stake in the tender.

As with all stages of PFI, the post-tender stage can be drawn out by client queries, post-tender changes and negotiations which may involve extensive tender revisions, particularly where a project is over budget.

Another highly important post-tender operation is feedback from the tender in terms of market prices and also lessons to be learned from the tender. The preparation of a PFI tender is a major project in itself and any lessons that can be learned to make the preparation of future tenders more efficient are very valuable

11.4   Value management

The client's requirements are reviewed and ranked in terms of importance. On a commercial contract the ranking may be:

- fit in the maximum number of employees

- finish the working areas as soon as possible

- keep the contract costs as low as possible

- high quality exciting design in customer areas

- good facilities for employees

- quality and space for directors' areas and the board room.

Design solutions are then reviewed, costed and ranked against the same list to establish whether the client's funds are being spent on the basis of the same priorities. If for instance the cost per square metre of the directors' areas is above that of the customer areas the designs should probably be changed so as to reflect the client's wishes. This process is by no means straightforward, nonetheless, it can identify areas of possible poor value to be re-examined and alternative design solutions considered.

On 2ST, D&B, management and construction management contracts the contractor joins the consultant team at an early stage and can use his expertise to contribute to value engineering. First, advice given on cost and procurement times will assist in the assessment of the existing design; second, the contractor can use his knowledge to research alternative designs and suppliers and recommend better value solutions. Value management is in a sense similar to risk management in that perceptions of priority and best value are partly subjective and also dependent on different people's knowledge and expertise. What may be seen as essential and very good value by the client's architect may be seen as costly and unnecessary by the PQS. For this reason, risk management workshops where design issues are considered in an informal meeting can be more valuable and productive than an attempt at formal design/cost/value assessments. The main objective should be to agree which design solutions are commonly agreed to be showing poor value and how to value engineer design and cost changes that would free up funds to enhance other areas or show a cost saving to the client.

Another aspect to be considered by the contractor is the project planning and the master programme. A procurement programme review must be carried out alongside design and cost reviews to ensure that the plant or materials specified will be available on site in time to meet the master programme. Further alternatives may have to be considered if cheaper materials and plant are on longer delivery periods. Alternatively, tactical short-term project planning alternatives may be needed to allow for extended procurement periods for better value items.

When firm costs have been established for elements of the design, whether under 2ST, D&B, construction management, management contracting or CLS contracts, divergences from the cost plan can be identified that may require a further stage of value engineering which is carried out by the estimator with sub-contractors or trade contractors. Projects which are over budget can be 'rescued' by the contractor working with his supply chain, the PQS, the design team and the client.

12 Computerisation and e-commerce

12.1   Computerisation

To get the best out of information technology (IT) systems, the focus from the very start of the tender process needs to be on what a software package can deliver. Any system that promises future benefits for large investments of time with no current benefit is poor value and not an effective tool. Open systems that will exchange information with a wide range of standard software produce more long-term benefits than closed systems. These are not compatible with software from other companies and will limit the involvement of the company in e-commerce. IT solutions can be fundamental in producing an integrated system for companies that link information from different departments and activities. For instance, cost planning by an estimating department can be adapted and used by the contracts department for valuations and in turn used by an accounts department for reporting contract profitability and cost divergences.

Large and complex systems have inherent risks of software and hardware faults and human errors which can have a much greater impact than with simple solutions. The best IT solutions are simple and provide the bare necessities required by the operator or company but have the potential for future expansion and improvement should the need arise. No IT solution will be 'future proof' but if it is adaptable at reasonable cost it can be developed to meet changing needs.

One of the great strengths of IT is the way in which communication can be improved. As referred to above, company internal IT systems can be used to share and develop cost information generated by the estimating process across other departments. With the use of external electronic links closer contact up and down the supply chain can be achieved. This could be on sharing information, be it on costs, design or programme, with clients, designers, contractors, sub-contractors and suppliers. The speed with which the transfer of information is possible can be highly beneficial in the current contracting environment. For example, systems exist that will take a hand-drawn concept of, say, an airport toilet from the architect, and develop that into computer-aided design (CAD) working drawings in the contractor's design department. These can be subsequently sent to a manufacturer (perhaps overseas) for a CAD-driven machine to produce the laminated wall and cubicle panels for delivery and site fixing.

Three-dimensional modelling gives contractors the ability to carry out highly sophisticated design co-ordination exercises using the CAD output from all the designers on a project. For instance, pinch points and clashes in service routes can be identified and designs adapted to minimise site installation problems, which is particularly valuable when off-site prefabrication is to be carried out. Computer-based planning and scheduling systems can be linked to a cost database to provide detailed budgets and cash flow forecasts for projects.

There is a great deal of development taking place in construction IT practice, much of it organically without best practice guidelines. What we see now is work in progress with many new developments which may affect the estimating process. The software used now for costing, planning and presentation is relatively well advanced; however, procedures for sharing CAD and other information needs industry standards to improve document control. The ability of CAD systems to produce quantities with associated costs has been available for a while but their reliability depends on the accuracy of the data recorded on the CAD drawings. Certain sectors of the construction industry have been able to utilise this effectively in their estimating, but this requires consistency of costs for items between projects as can be demonstrated in the manufacturing industry. Here the production environment is much more controlled reducing the variances considerably.

12.2   e-Commerce

Since the previous edition of this book, IT has moved on in leaps and bounds. As the construction industry makes more investment in digital technology and more reliance is placed on electronic communication, e-commerce will become prevalent in the working arena. e-Commerce in the context of estimating and the pre-construction process is a way of managing the entire contract letting process electronically.

Computer technology has now advanced to the stage where it is possible to receive, store, separate and re-transmit tender documents to all members of the design and construction team. Provided the organisation has the appropriate hardware and software, electronic transmission of letters, drawings, specifications and other tender documents can all be achieved electronically.

The UK government is one of many key clients spearheading the adoption of e-commerce. A number of initiatives and pilot schemes are under way. These are aimed at developing the use of e-commerce while addressing the issues of security, ease of use, compatibility and legality.

What is e-tendering?

In general e-tendering in construction refers to the process of issuing electronic tender documentation to main contractors, sub-contractors and industry suppliers and receiving their responses electronically. However, further benefits for the estimator include software packages allowing accurate cost planning and storage of historical data, pricing of bills of quantities using a price database and post-contract financial management systems.

Recent developments in e-tendering software, along with the spread of high quality low-cost connections such as broadband Internet access, have meant that e-tendering is slowly becoming a reality in the construction sector. However, as with all IT-driven change, the transition from paper-based to e-tendering is not a simple one. The potential benefits of e-tendering include a reduction in the amount of paper used in the process, a reduction in the administration work required to issue and respond to tenders, a higher level of accuracy and speed and the cost savings resulting from these benefits.

12.3   e-Bidding processes

The process

All the documentation is distributed to tenderers electronically or via a secure web-based system. The client's team avoids the need for collating all the paperwork and sending it by courier. In due course the tenderers return their bids the same way.

During the tender period, updates and queries are exchanged through the same system. This means that all the key information is at the fingertips of those involved. The client has everything needed to assess the bids, and the software will generally contain tools which help in comparison of the tenders.

If there is bill of quantities with the tender documentation, it is usual for this to have been created by a quantity surveying firm using specialist software. Once the bill of quantities has been approved, it becomes part of the tender package, along with any necessary drawings needed by tenderers to compute their final tender price bids. The term 'electronic' tendering refers to the process of distributing these tender packages in digital rather than paper format.

In the current market, it is advisable to insist that a paper copy of the tender documents is provided by the employer's design team or the quantity surveyor to ensure that there are no errors in the provision of the electronic documentation.

e-Tendering can be used for all values of tenders, although it is primarily used for higher value contracts at present.

The working environment

Equipment requirements

This chapter does not consider the particular specification for the computer and associated equipment and software as the developments in this field are changing on such a regular basis that it would be out of date by the time this book reaches the reader.

Estimators joining organisations will need to evaluate the level of equipment within the firm and establish whether this is sufficiently complete to cope with large files that transmit tender documents and drawings. The estimator will also need to establish whether equipment and software are able to separate the documents for onward transmission to other parties to the tendering process.

With some major projects the employer's design team will set up a central server (managed by an independent firm or one member of the professional team) that holds all of the current documents pertaining to a project. The tenderer can access this information for the purpose of tender and download any of the relevant tender documents for the purpose of tender preparation. In most instances, the software for a central server is held on the system and is freely available to tenderers. The server is protected by password access.

Careful consideration is required to ensure that the control of documents is managed effectively; therefore when e-commerce is implemented robust procedures must be operated.

Applications

Computer-aided estimating

Over the past few years, there has been a consolidation of the estimating software packages available. This has led to more investment by the providers and means that the usability and the benefits are of great value to the estimating function.

The following is a résumé of how a modern estimating software package works.

Bills of quantities

Bills of quantities are normally loaded into the system from scanners, e-mail, ASCII, CITE or Excel files. Where smaller jobs are to be estimated on the system there will normally be a manual rapid entry system.

As well as pricing, the software normally handles the administration of a tender. When creating a new job you will be invited to complete a tender register for the project.

Analytical estimating

Analytical estimating has become more and more simple and straightforward over the years. Even the most complex bill items can easily be priced using a whole range of pricing techniques. Once again flexibility is the key; each estimator is likely to approach the pricing of the same item in a different way. Some work with outputs, some with constants and some will work out a total and divide by the bill quantity.

A good system should give the estimator total flexibility. For some trades you may wish to price items yourself as well as sending out for sub-contract prices, deciding which rate to use later on. All of these methods and more are now available.

Cost planning

The continued growth in budget pricing, 2ST and other new forms of contract means that software has been introduced to deal with these procurement routes.

Once an outline budget price is mentioned to a client it can very quickly become, in his mind, an estimate and then a maximum fixed price tender figure! For this reason it is as well to ensure that one's cost planning is based on sound information, preferably your own information based on accurate historical data. In general terms the operator's software will retrieve historical costs which will enable a fast and accurate cost plan to be compiled.

Project planning The computer can greatly simplify project planning and control because:

- the mathematical analysis is carried out automatically

- quickly finished documents such as cascade diagrams, bar charts, histograms, graphs and PDM networks are produced directly on computer printouts

- detailed resource analysis, including financial forecasting and cost reporting, can be easily carried out

- the network calculations and outputs can be updated rapidly and cheaply

- the inclusion of networks and other systems into an integrated management control system is very easily accomplished.

However, while most software products will enable the user to prepare a visually attractive diary of dates, not all software is suitable for dynamic time control. Some software actually inhibits effective time control by facilitating an absence of consistency in activity identification, limiting the number and type of logical connections or the ability to identify the effect of progress achieved on completion dates.

For effective time control it is essential that updated reviewed and revised project programmes have consistent activity identifiers and that updates are carried out from a straight data date-line with all works actually done to the left of the date line and all work to be done to the right of it. Any software that permits the user to simply monitor progress by attributing a degree of completion to planned activities without changing their duration and/or timing to identify the effect of progress achieved on uncompleted work is useless for time management.

Computerisation and e-commerce

The facility within the software for breaking down a project into sub-projects with independent and multiple access to each may also be important in large projects that require a team of managing project schedulers.

Receipt of information

The registering of tender documents received should occur whether the information is received electronically or not. See Chapter 3.

Records

Keeping track of enquiries

The software package will normally include an enquiry tracking system that compiles a list of all of the enquiries produced for a job. For each enquiry it records what was sent, including drawing lists, etc., when it was sent, when it is due back in, who it was sent to, and contact details including telephone numbers, fax numbers, e-mail addresses and names.

All of this information is held in a database which allows a search to be made by trade or date or any other part of the stored information. As replies to enquiries come in they can be ticked off as returned. Normally there is a notes facility for recording any particular details or contact made.

Comparison of quotations

Comparison sheets will normally be created automatically for each trade. The system sets up lists of items for each trade along with the names of the companies to which the enquiry was sent. Non-respondents can be deleted and columns can be added for any unsolicited tenders. Modern systems are capable of handling tenders in whatever way they are compiled whether items have been individually rated, or priced as page totals only or lump sums.

Missing rates can be inserted with lowest, highest, average rates or with your own analytical rate directly from the bill if you have priced it. You can normally add on for attendances, preliminaries, number of visits, etc. Good systems can also cope with main contractor's discount, fixed price terms and any other conceivable difference. Once you are happy that you have compared as close to like-for-like as possible, the software will rank the quotes, lowest first.

You then choose the one you want to use and the system will update the bill automatically. If you have already priced the items in question yourself the system will ask which elements of your own analytical build-up you want to replace, labour only, labour and plant or any other combination. If a preferable late quote arrives the system will swap one for the other in an instant.

Final adjustments

There are many facilities in modern estimating packages which will help you to manipulate your net estimate to produce a final tender. Resource levelling will allow you to adjust total resource quantities to take account of minimum hire periods, non-productive time, wastage, buying quantities, etc.

Lump sum and percentage adjustments can be made item by item, over ranges of items, pages or sections. Global adjustments can be made for overheads and profit. In addition, it is possible to link the software to an adjudication sheet created to your specification.

The future

Software packages exist that purport to be able to measure quantities and price projects from a standard library of descriptions and prices. This type of software is in a very early stage of development and this approach to tendering is not to be encouraged at this time. The risk that a quantity, description or price is misinterpreted by the software is very high.

Computerisation and e-commerce

The potential does exist for sub-contractors and suppliers to maintain comprehensive websites that hold the firm's current prices, thereby avoiding the need for parties in the tendering process to exchange bulky documents. It could mean that only the customised works need to be carefully considered and priced by the contractor.

Progress records

Effective records of resources used and productivity achieved are the skeleton on which effective project planning is built for the tendered works and delay and disruption claims are supported for changed works. Unless good records are kept of what was done, where, by whom, when, and how that departed from what was tendered for, no claim for additional costs or time can be proved.

Paper-based records cannot be searched, retrieved, compared or accumulated. Accordingly, contractors should consider either having paper records rekeyed at weekly intervals into a database or having the data entered directly into a database that will facilitate the preparation of fast, accurate, progress reporting.

Microsoft Access is a good relational database that will easily facilitate the keeping and retrieval of progress records. Essentially, the records that should be kept and reviewed are:

- For the tendered works

    - date

    - day of week

    - name of resource

    - trade

    - rank or grade of resource

    - task identified by reference to programme

    - activity worked on by reference to programme

    - area worked on by reference to programme.

- For work that is different from that tendered for

    - an allocated activity identifier

    - architect's instruction or other contract document instruction the work

    - task description

    - labour allocation

    - plant allocation

    - materials allocation

    - date of work carried out

    - predecessor logic

    - successor logic.

References and further resources

Aqua Group (1999) *Tenders and Contracts for Building*, 3rd edition, Blackwell.

Aqua Group (2003) *Pre-Contract Practice for the Building Team*, Mark Hackett and Ian Robinson (Eds), Blackwell.

Ashworth, A. (2004) *Cost Studies of Buildings*, 4th edition, Pearson Prentice Hall.

BCIS (2008) *Guide to Estimating for Small Works 2008*, Building Cost Estimation Service.

BEC/RICS (1998) *Standard Method of Measurement of Building Works*, 7th edition (revised).

Brook, M. (2008) *Estimating and Tendering for Construction Work*, 4th edition, Butterworth Heinemann.

Buchan, R.D., Fleming, E. and Grant, F.E.K. (2003) *Estimating for Builders and Surveyors*, 2nd edition, Butterworth Heinemann.

Chartered Institute of Building (2009) *Code of Practice for Project Management*, 4th edition.

Chartered Institute of Building (1997) *Code of Practice for Estimating*, 6th edition.

Civil Engineering Standard Method of Measurement, 3rd edition (1991), Thomas Telford.

Cook, A.E. (1991) *Construction Tendering: Theory and Practice,* Batsford.

Farrow, J.J. (1984) *Tendering – An Applied Science*, 2nd edition, The Chartered Institute of Building.

Fatzinger, J., Dupuiss, V., Mudial, D. and Hall, G. (2003) *Basic Estimating for Construction*, Prentice Hall.

Franks, J. (1998) *Building Procurement Systems: A Client's Guide*, 3rd edition, The Chartered Institute of Building.

Holroyd, T. (2000) *Principles of Estimating*, Thomas Telford.

JCT Practice Note 6 Series 2 (2002) *Main Contract Tendering*.

Jones, J. (1986) *Handbook of Construction Contracting: Estimating, Bidding, Scheduling* (Vol. 2), Craftsman.

Kwakye, A.A. (1994) *Understanding Tendering and Estimating*, Gower.

Langmaid, J. (2003) *Estimating, Getting Value from Function*, BSRIA.

Latham, Sir Michael (1994) *Constructing the Team* (joint review of procurement and contractual arrangements in the United Kingdom construction industry), HMSO.

McCaffer, R. and Baldwin, A. (1991) *Estimating and Tendering for Civil Engineering Works*, 2nd edition, Blackwell.

Peterson, S. (2007) *Construction Estimating using Excel*, Prentice Hall.

Pratt, D. (2004) *Fundamentals of Construction Estimating*, 2nd edition, Thomson.

Sher, W. (1997) *Computer-aided Estimating: A Guide to Good Practice*, Addison Wesley Longman.

Skitmore, R.M. (1989) *Contract Bidding in Construction: Strategic Management and Modelling*, Longman Scientific & Technical.

Smith, A.J. (1995) *Estimating, Tendering and Bidding for Construction*, Macmillan.

Smith, R.C. (1986) *Estimating and Tendering for Building Work*, Longman.

Further resources

Spon's Price Books (Taylor & Francis):

- *Spon's Architect's and Builder's Price Book 2009.*
- *Spon's Civil Engineering and Highway Works Price Book 2009.*
- *Spon's Mechanical and Electrical Services Price Book 2009.*
- *Spon's External Works and Landscape Price Book 2009.*

CIOB Technical Information Papers and other relevant publications (1978–1991):

- Paper No. 7 Potter, D. and Scions, D. (1982) Computer aided estimating.
- Paper No. 9 Holes, L.G. and Thomas, R.S. (1984) Pricing drainage and external works.
- Paper No. 11 Harrison, R.S. (1982) Practicalities of computer aided estimating.
- Paper No. 15 Skoyles, E.R. (1982) Waste and the Estimator.
- Paper No. 37 Harrison, R.S. (1984) Pricing drainage and external works.
- Paper No. 39 Braid, S.R. (1984) Importance of estimating feedback.
- Paper No. 59 Uprichard, D.C. (1986) Computerised standard networks in tender planning.
- Paper No. 64 Ashworth, A. (1986) Cost models – their history, development and appraisal.
- Paper No. 65 Ashworth, A. and Elliott, D.A. (1986) Price books and schedules of rates.
- Paper No. 75 Harrison, R.S. (1987) Managing the estimating function.
- Paper No. 77 Ashworth, A. (1987) General and other attendance provided for subcontractors.
- Paper No. 81 Ashworth, A. (1987) The computer and the estimator.
- Paper No. 97 Brook, M.D. (1988) The use of spreadsheets in estimating.
- Paper No. 113 Senior, G. (1990) Risk and uncertainty in lumpsum contracts.
- Paper No. 114 Emsley, M.W. and Harris, F.C. (1990) Methods and rates for structural steel erection.
- Paper No. 120 Cook, A.E. (1990) The cost of preparing tendering tenders for fixed priced contracts.
- Paper No. 127 Harris, F. and McCaffer, R. (1991) The management of contractor's plant.
- Paper No. 128 Price, A.D.F. (1991) Measurement of construction productivity: concrete gangs.
- Paper No. 131 Brook, M.D. (1991) Safety considerations on tendering – management's responsibility.
- The Chartered Institute of Building (1978) *Information Required before Estimating: a code of procedure supplementing the Code of estimating practice*.
- The Chartered Institute of Building (1981) *The Practice of Estimating* (compiled and edited by P. Harlow).
- The Chartered Institute of Building (1983) *Code of Estimating Practice*, 5th edition.
- The Chartered Institute of Building (1987) *Code of Estimating Practice*, Supplement No. 1: *Refurbishment and Modernisation*.
- The Chartered Institute of Building (1988) *Code of Estimating Practice*, Supplement No. 2: *Design and Build*.
- The Chartered Institute of Building (1989) *Code of Estimating Practice*, Supplement No. 3: *Management Contracting*.

CIOB Technical Information Papers and other relevant publications (1992–):

- Paper No. 2 Massey, W.B. (1992) Subcontracts during the tender period – an estimator's view.
- Paper No. 11 Hardy, T.J. (1992) Germany – a challenge for estimators.
- Paper No. 16 Milne, M. (1993) Contracts under seal and performance bonds.
- Paper No. 17 Young, B.A. (1993) A professional approach to tender presentations in the construction industry.
- Paper No. 19 Harrison, R.S. (1993) The transfer of information between estimating and other functions in a contractors' organisation – or the case for going round in circles.
- Paper No. 23 Emsley, M.W. and Harris, F.C. (1993) Methods and rates for precast concrete erection.
- Paper No. 32 Morris, R. (1994) Negotiation in construction.
- Paper No. 33 Harrison, R.S. (1994) Operational estimating.
- Paper No. 49 Borrie, D. (1995) Procurement in France.
- Paper No. 54 Pokora, J. and Hastings, C. (1995) Building partnerships.
- Paper No. 55 Sher, W. (1995) Classification and coding data for computer aided estimating systems.
- The Chartered Institute of Building (1993) *Code of Estimating Practice*, Supplement No. 4: *Post-tender Use of Estimating Information*.

Index